Lightweight Hardware Security and Physically Unclonable Functions

Kasem Khalil • Haytham Idriss • Tarek Idriss •
Magdy Bayoumi

Lightweight Hardware Security and Physically Unclonable Functions

Improving Security of Constrained IoT Devices

 Springer

Kasem Khalil
University of Mississippi
Oxford, MS, USA

Tarek Idriss
Western Washington University
Bellingham, WA, USA

Haytham Idriss
Purdue University Fort Wayne
Fort Wayne, IN, USA

Magdy Bayoumi
University of Louisiana at Lafayette
Lafayette, LA, USA

ISBN 978-3-031-76330-4   ISBN 978-3-031-76328-1   (eBook)
https://doi.org/10.1007/978-3-031-76328-1

This Springer imprint is published by the registered company Springer Nature Switzerland AG
The registered company address is: Gewerbestrasse 11, 6330 Cham, Switzerland

If disposing of this product, please recycle the paper.

Preface

The rapid growth of the Internet of Things (IoT) and the increasing deployment of resource-constrained devices in critical applications such as healthcare, supply chain management, and smart cities have brought unprecedented convenience and connectivity. However, these benefits come with significant security challenges. IoT devices often operate with limited processing power, memory, and energy resources, making them particularly vulnerable to attacks. Traditional cryptographic security measures are often too costly in terms of power and computational requirements, leaving these devices exposed to a range of threats. This book is a comprehensive exploration of the emerging field of lightweight hardware security with a focus on Physically Unclonable Functions (PUFs). PUFs represent a new class of hardware security primitives that leverage inherent manufacturing variations in silicon to provide unique, unpredictable responses to external challenges. Unlike conventional cryptographic methods, PUFs do not require key storage, making them highly attractive for constrained environments where space, power, and security are critical concerns. This book is structured to provide a detailed understanding of the security landscape for resource-constrained devices and the role of PUFs in enhancing security. The early chapters introduce the reader to the fundamental security challenges of IoT devices, particularly Radio Frequency Identification (RFID) systems, which are some of the most constrained yet widely used components in IoT networks. From the vulnerabilities and architectural constraints of these systems, the discussion shifts to how lightweight security protocols can mitigate risks but still face limitations against sophisticated attackers. The core of this book delves into the design, implementation, and application of PUFs, highlighting their potential as a lightweight security solution that does not rely on traditional key management. The chapters explore various types of PUFs, their operational principles, and the metrics that define their effectiveness, such as inter-chip and intra-chip Hamming distances. It provides a detailed evaluation of PUF performance metrics and introduces advanced designs and lightweight security protocols tailored for resource-constrained environments. The book also differentiates between strong and weak PUFs, each suitable for different security applications. It provides a guideline for choosing the most suitable design for each application, from device authentication to secure key generation. PUFs are not without their challenges. Managing challenge-response pairs, ensuring reliable operation under varying conditions, and protecting against modeling attacks are crucial areas of research. This book discusses these challenges and offers insights into current strategies and future directions in the field. This book aims to serve as a valuable resource for researchers, engineers, and practitioners working in the field of hardware security. It combines theoretical foundations with practical insights, offering readers a balanced perspective on the opportunities and challenges of integrating PUFs into real-world security solutions.

Oxford, MS, USA   Kasem Khalil
Fort Wayne, IN, USA   Haytham Idriss
Bellingham, WA, USA   Tarek Idriss
Lafayette, LA, USA   Magdy Bayoumi

Acknowledgment

The authors would like to acknowledge the VLSI Group, Center for Advanced Computer Studies (CACS), and the Electrical and Computer Engineering Department at the University of Louisiana at Lafayette. The research environment fostered by this group has been a source of invaluable inspiration and insight. The collective expertise, dedication, and intellectual curiosity of the faculty, researchers, and students have played a pivotal role in shaping the ideas and concepts explored in this book. We are grateful for the countless discussions, feedback, and collaboration that have enriched the content of this book.

We would also like to extend a special acknowledgment to Sonia Akter for her contribution, support, and help throughout this project.

About the Book

This book delves deep into the realm of lightweight hardware security, focusing on Physically Unclonable Functions (PUFs). As security becomes increasingly important in modern hardware design, especially in resource-constrained environments like the Internet of Things (IoT), PUFs have emerged as a groundbreaking solution. The book explores the fundamental principles of PUFs, their applications in device authentication and cryptographic key generation, and their ability to resist cloning. It also covers common security challenges, such as side-channel attacks and threats based on machine learning. Additionally, it delves into advanced PUF designs, performance evaluation, and the development of lightweight PUF-based protocols for secure communications. By providing practical solutions to enhance the reliability and security of hardware devices, the book offers a comprehensive framework for implementing efficient, scalable, and secure systems in both academic research and industrial applications.

Contents

Chapter 1
Introduction and Background

Physical Unclonable Functions (PUFs) have emerged as a vital component in the field of hardware security, providing robust solutions for device authentication and cryptographic key generation. A PUF exploits the inherent physical variations that occur naturally during the manufacturing process of integrated circuits (ICs). These variations are uncontrollable and unique to each IC, making it virtually impossible to replicate or clone. As a result, PUFs can generate unique responses or signatures that serve as hardware fingerprints, which are used to verify the identity of devices in a secure manner. PUFs operate by applying a challenge to the hardware and observing the corresponding response. Due to the unique physical properties of the IC, the response is both unpredictable and consistent, providing a reliable basis for security applications. PUFs have found applications in a variety of fields, including secure communications, digital rights management, and the Internet of Things (IoT).

In the context of the increasing number of resource-constrained devices, particularly in the IoT ecosystem, lightweight security has become an essential consideration. Lightweight security refers to cryptographic protocols and security mechanisms that are optimized for devices with limited computational power, memory, and energy resources. These protocols ensure that even devices with stringent resource constraints can achieve an acceptable level of security without compromising performance or battery life. PUFs play a crucial role in lightweight security due to their minimal overhead. Traditional cryptographic methods often require substantial computational resources, which are not feasible for many low-power devices. PUFs, on the other hand, offer a hardware-based solution that is both efficient and secure, making them ideal for lightweight security applications.

Hardware security encompasses a broad range of practices and technologies designed to protect hardware devices from unauthorized access, tampering, and other malicious activities. It involves securing the physical components of a system and the data and processes that operate within these components. As hardware becomes increasingly integrated into critical infrastructure and everyday devices, the importance of robust hardware security mechanisms has grown significantly.

K. Khalil et al., *Lightweight Hardware Security and Physically Unclonable Functions*, https://doi.org/10.1007/978-3-031-76328-1_1

PUFs are a key innovation in hardware security, providing a tamper-evident and resistant solution. Unlike traditional security mechanisms that rely on software or external encryption keys, PUFs embed the security directly into the physical substrate of the hardware. This makes them inherently resistant to a wide range of attacks, including invasive and noninvasive tampering efforts. The integration of PUFs into hardware security frameworks helps ensure that devices can be securely authenticated and that cryptographic keys can be generated and stored securely within the hardware itself. This approach mitigates the risk of key extraction and other forms of attack that target software-based security mechanisms.

1.1 The Need for a Revolutionary Lightweight Security Suite

Our world has become a smarter, more connected world where various devices and systems are utilized to enhance our daily lives. We now live in the world of Internet of Things or IoT rather than the Internet of People. Smart, connected devices now help us monitor water levels or simply find the nearest parking spot. Others autonomously manage lighting systems or irrigation systems in agricultural facilities. In a report by British Intelligence (BI), it is projected that, by 2020, there will be around 24 billion smart objects connected to the Internet of Things (IoT), which is more than six times the world's projected population at the time. The smart objects in IoT will pervade all aspects of our daily lives and fundamentally alter the way we interact with our physical environment, thereby revolutionizing a number of application domains such as telemetry, healthcare, home automation, energy conservation, security, wearable computing, asset tracking, maintenance of public infrastructure, and many others [57, 77, 79–81]. However, with these quality of life improvements, new cyberthreats have risen. In recent years, there has been an unprecedented level of security breaches in various systems across the world.

Multiple security assessments of the IoT security scene have shown an abysmal level of security flaws when in connected devices. A research report by Hewlett Packard Enterprise [113] has shown that more than 70% of connected devices have significant security vulnerabilities, including personal information leaks, unencrypted connection use, and lack of secure passwords. Another IoT security assessment effort [1] was performed by Kaspersky Labs and found vulnerabilities in critical applications such as SCADA (Supervisory Control and Data Acquisition) systems and traffic controllers. More recently, IoT security issues have been showing in the news headlines as privacy issues have been identified in wearable fitness trackers [64], while medical heart implants were recalled due to cybersecurity concerns [101]. Indeed, it has become clear that IoT security is the most important challenge hindering the IoT growth.

There have been various research efforts aiming to tackle the security challenges hindering the wide adoption of IoT. However, with security challenges present in all layers of the IoT and incorporating various new technologies, it is very difficult for researchers to address all challenges without a novel IoT security framework. A framework that can categorize the devices and technologies of the IoT according their security threats and security requirements would allow a directed research effort that can tackle security threats with a surgical precision.

Developing such a framework would be a top research priority. Nonetheless, IoT security needs to be established on a device-to-device basis until such a framework is available. The security threats faced by the endpoint devices need to be ultimately addressed by the device hardware and application developers. The threats and vulnerabilities must be examined carefully, while suggested solutions need to be within the computational capacity of each device.

With the increasingly large number of successful cyberattacks on systems on devices across the globe, it is important to examine the factors behind the increased vulnerabilities of our systems. We recognize four major causes for the increase in system vulnerabilities:

1. Negligence: Unfortunately, manufacturer negligence has been the prime factor in increased device vulnerabilities. Devices were shipped with generic "admin" username and password. Access modes for debug were not properly closed to end users, while others had passwords that were easy to figure out.
2. Lack of Product Updates and Support: Suppliers often do not provide a stream-lined update process (no autoupdate or link to Internet). Products may also have short-lived support, with vendors no longer providing updates for older devices.
3. New Protocol Vulnerability: New lightweight communication protocols were suddenly deployed "en masse". Many vulnerabilities were discovered and required patching. Moreover, developers seemed to lack the proper expertise and knowledge of new protocols leading to faulty or insecure implementations.
4. Constrained Devices: The deployment of highly constrained devices that cannot employ standard cryptography protocols leads to a large number of security breaches. Due to the lack of a reliable lightweight security solution, these products were shipped with make-shift security that fails against dedicated attackers.

Most of the security vulnerabilities recognized can be mitigated using good design practice and vigilance. However, the challenge facing constrained devices would require a new security solution that has lightweight computation. In our research, we tackle the security challenges facing the most vulnerable devices, which are constrained devices that lack the ability to deploy reliable cryptographic primitives due to the high constraints on power consumption, implementation area, or device cost. Establishing reliable security for such constrained devices has been an ongoing challenge for more than a decade. Standard cryptographic solutions that provide provable security have prohibitive area and power demands for many applications like RFIDs, medical implants, or smart cards. For instance, low-cost RFID tags can only use 3k–5k logic gates for security functions, as reported in [30] and [137], while public cryptography algorithm implementations, which are crucial for reliable key exchange [13, 20], can use between 12k and 22k logic gates [97].

Therefore, we had to look into new technologies for reliable, lightweight security as traditional approaches have repeatedly failed against dedicated attackers. PUFs, or Physically Unclonable Functions, are emerging hardware security primitives that offer a lightweight alternative to standard security. PUFs, however, have been found to be vulnerable to modeling attacks. In current literature, PUFs are yet to be secured reliably without the use of cryptographic functions. In this dissertation, we design and develop a full PUF-based security suite for constrained devices that has a lightweight implementation and utilizes no cryptographic functions. The security suite would be the first to offer unlimited mutual authentication and a cipher for secret message exchange for constrained devices.

1.2 Introduction to Cryptographic Ciphers

Cryptographic ciphers have been extensively used in the past few decades to secure communication and transactions. While ciphers are mathematically secure, faulty implementations can leave them vulnerable to side-channel attacks or leave back doors open to hackers. For proper implementations of cryptographic primitives, it is vital to understand the details of the algorithms utilized by secure cryptographic ciphers.

1.2.1 Introduction to Cryptography

Cryptography is the practice of securing information by transforming it into a coded format, ensuring that only authorized parties can understand its true meaning. Cryptography has been used since the earliest ancient civilization, with early examples dating back to about 2000 B.C., when nonstandard "secret" hieroglyphics were used in ancient Egypt [122]. Since then, cryptography has been used in one form or the other in many cultures that developed written languages.

As Fig. 1.1 illustrates, cryptography is one branch of cryptology with the other branch being cryptanalysis which is the science of extracting hidden information

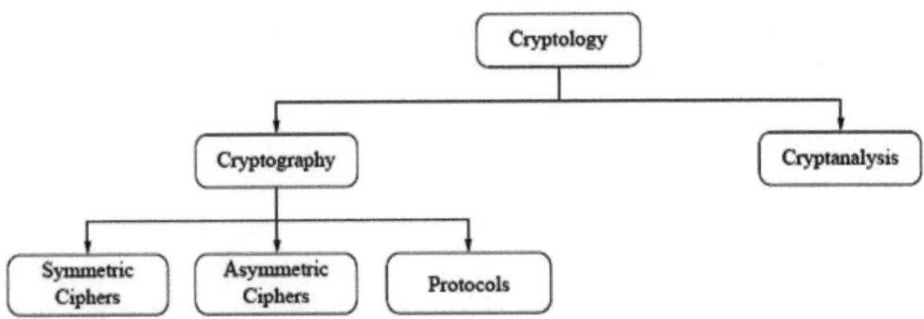

Fig. 1.1 Cryptography branches

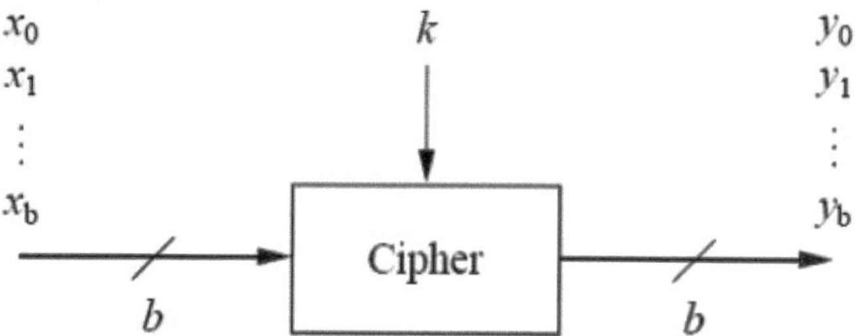

Fig. 1.2 Cipher used for encrypting plain text x into cipher text y

from secret texts. Cryptography on the other hand is divided into three parts: Symmetric Ciphers, Asymmetric Ciphers, and Protocols which are required to ensure that ciphers are used securely.

1.2.2 Cipher Security

A cryptographic cipher transforms a plain text x into a cipher text y that is unique to the key k, used as shown in Fig. 1.2. The security of the cipher does not lie in having neither the encryption or decryption algorithms hidden nor their implementations. According to Kerckhoffs' principle [24], a cryptosystem should be secure even if the attacker knows all details about the system, with the exception of the secret key. In particular, the system should be secure when the attacker knows the encryption and decryption algorithms. Ciphers that relied on hiding the algorithm or their implementation have failed repeatedly. An example of such ciphers is the CSS (Content Scramble System) [12] cipher which was used on DVDs (Digital Versatile Disks) to hide contents.

Claude Shannon has defined the core operations that need to be performed by a cipher to ensure its security:

1. Confusion is an encryption operation where the relationship between key and cipher text is obscured. A common element for achieving confusion is substitution.
2. Diffusion is an encryption operation where the influence of one plain text symbol is spread over many cipher text symbols with the goal of hiding statistical properties of the plain text.

Modern ciphers need to ensure that such operations are performed sufficiently in the algorithms to thwart different kinds of attacks.

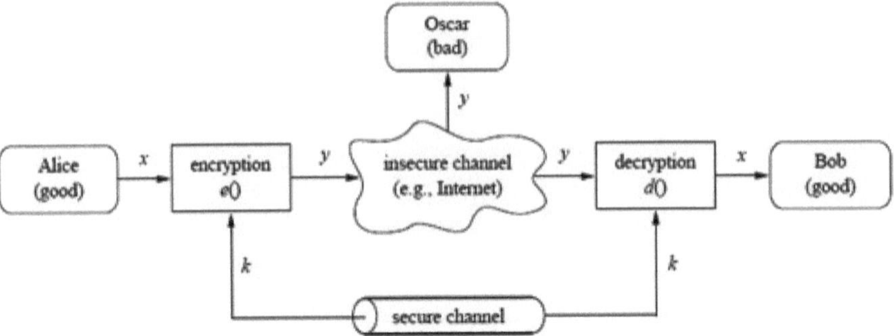

Fig. 1.3 Symmetric encryption

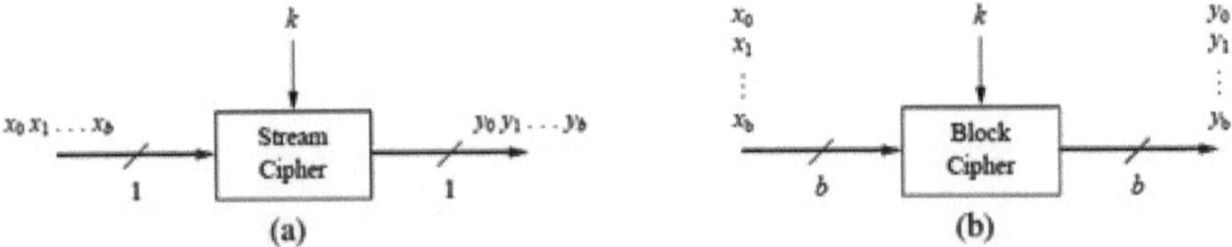

Fig. 1.4 Stream cipher versus block cipher

1.2.3 Symmetric Ciphers

In a symmetric cipher, the key that is used for encryption is used for decryption as well. The secret key needs to be transferred using a secure channel to ensure that untrusted parties are unable to decrypt the cipher text (Fig. 1.3). Symmetric ciphers can be divided into two categories: stream ciphers and block ciphers. Stream ciphers encrypt data one bit at a time, while block ciphers encrypt data one block at a time, as Fig. 1.4 shows. While this seems like a trivial distinction, there is a rather significant difference when it comes to the algorithms used in each.

1.2.3.1 Stream Ciphers

In a stream cipher, data is encrypted one bit at a time. A key is used in the generation of an unpredictable bit stream that is XOR-ed one bit at a time with the plain text bit. Aside from the key stream generation, the operation is rather a very simple binary XOR operation. This makes the stream ciphers more appealing for resource-constrained environments.

The security of stream ciphers lies entirely in the key stream generation. If a True Random Number Generator (TRNG) is used, the cipher becomes unbreakable. A stream cipher that uses a TRNG as its key stream generation is known as a One Time Pad (OTP) since the key stream is only used once. In an OTP cipher, the key length is the key stream itself since there is no way to regenerate the key stream. This makes it highly impractical since the key that needs to be transferred securely is of the same size as the data itself. Still, it is known that such ciphers were used before to establish secure diplomatic channels due to their unbreakable security. Security holes that were discovered were often the result of reusing portions of the key stream instead of using a true random stream. Currently, stream ciphers are less popular than block ciphers in most domains as block ciphers are more thoroughly studied and, hence, highly trusted. With the advent of the computational capacities of devices, computers are easily able to deploy block ciphers.

This takes us to the advantage of stream ciphers. Stream ciphers generally require fewer resources, e.g., code size or chip area, for implementation than block ciphers making them attractive for use in constrained environments such as cell phones. The entire security of a stream cipher lies in its key stream generation. Since true number generators cannot be used as there is no way to regenerate the key stream other than recording it entirely, systems usually deploy what is known as Pseudorandom Number Generators (PRNGs).

Unlike TRNGs, PRNG output can be predicted when a large portion of the stream has been observed. This is why PRNGs are always a focus of research efforts that aim to introduce a less predictable and ultimately more secure stream. Left shift registers connected in series can generate a seemingly random stream of bits when feedback is added to the loop. The stream will have to loop through the possible register value combinations, and once it reaches its initial state it will repeat. The number of bits generated is directly related to the number of registers used $(2N - 1)$, where N is the number of registers.

The simple LSR in Fig. 1.5 is obviously not secure and needs a much larger number of registers with more complicated feedback loops. Such LSR was implemented in the stream cipher knows as Trivium [32] and is shown in Fig. 1.6.

While creating a PRNG with good randomness characteristics is a rather simple task, the requirements for a cryptographically secure pseudorandom number generator are far more demanding. For a pseudorandom number generator to be cryptographically secure, it should be sufficiently unpredictable even with the knowledge of the previously generated bit.

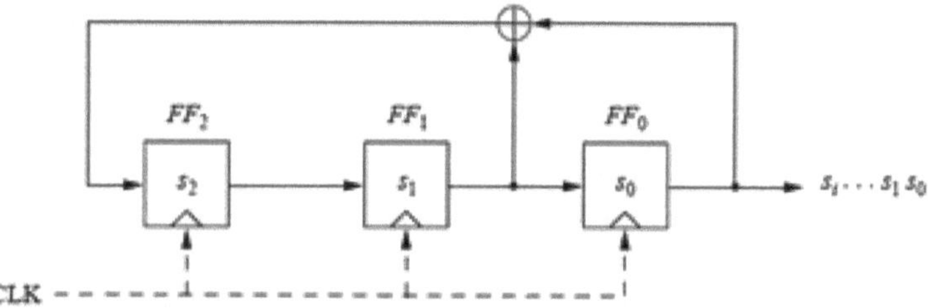

Fig. 1.5 LSR circuit architecture

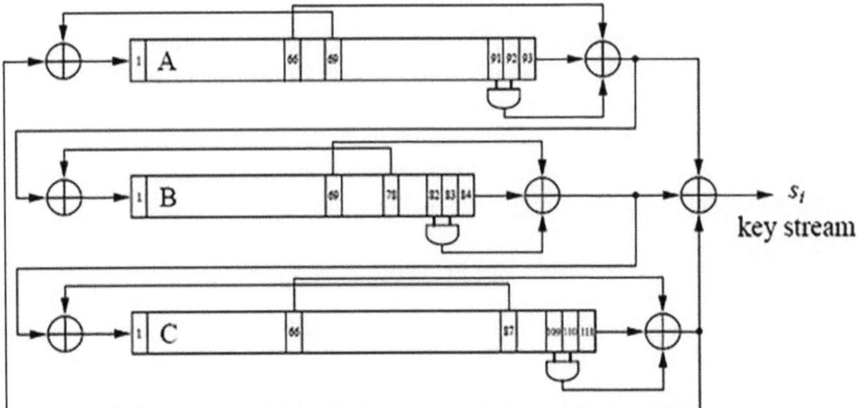

Fig. 1.6 Trivium LSR architecture

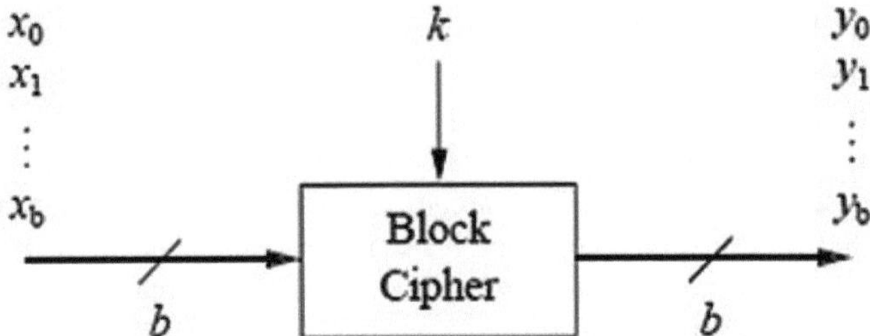

Fig. 1.7 Block cipher black box

1.2.4 Block Ciphers

Block ciphers are a fundamental component of modern cryptography, encrypting data in fixed-size blocks rather than individual bits. As illustrated in Fig. 1.7, a block cipher processes a block of data consisting of b bits and outputs an encrypted block of ciphertext of the same size. This process differs from stream ciphers, which encrypt data bit by bit using a PRNG. Instead, block ciphers employ sophisticated cryptographic techniques, such as the Feistel network and the SP (Substitution-Permutation) network, both of which are widely used and renowned for their security properties. These techniques, to be further explored in this section, ensure robust encryption by introducing complex permutations and substitutions within each block of data, making it resistant to various cryptographic attacks.

1.2.4.1 Feistel Networks

Feistel networks were first seen commercially in IBM's Lucifer cipher, designed by Horst Feistel and Don Coppersmith in 1973. A single round of encryption only

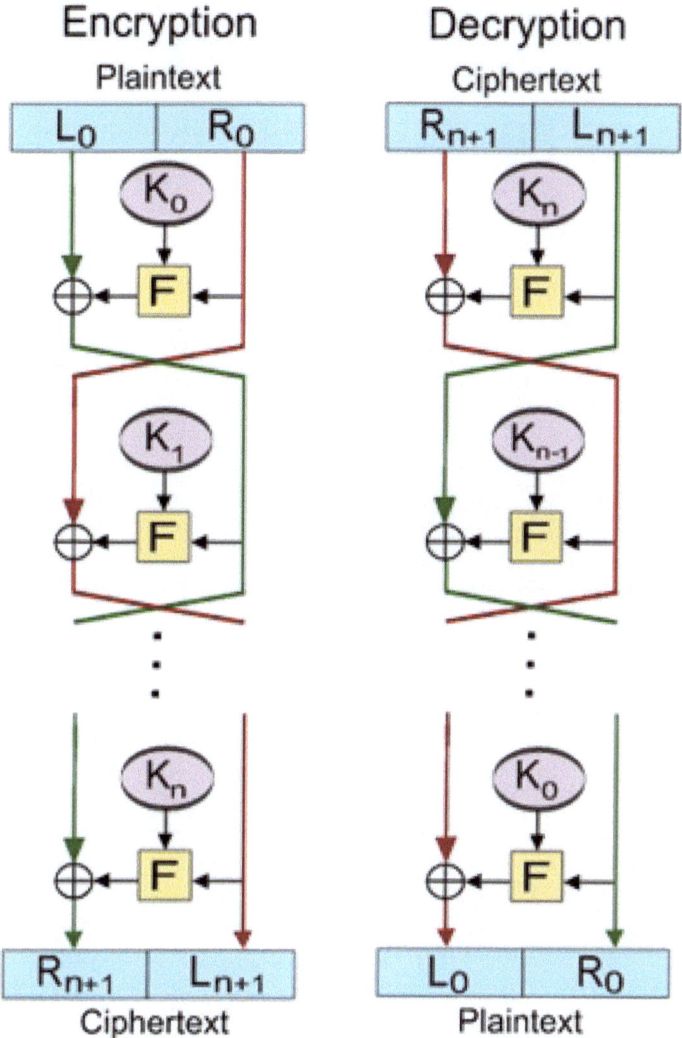

Fig. 1.8 Feistel network architecture

effectively encrypts half of the block while rotating the halves for the next round. By rotating the block halves and using the key at every round, both the confusion and diffusion properties are introduced.

Figure 1.8 shows the architecture of a basic Feistel network. Both the encryption and decryption networks are very similar, and in some cases they can be identical requiring only the reversal of the key scheduling. This makes the network quite efficient to implement in hardware when the size is of concern.

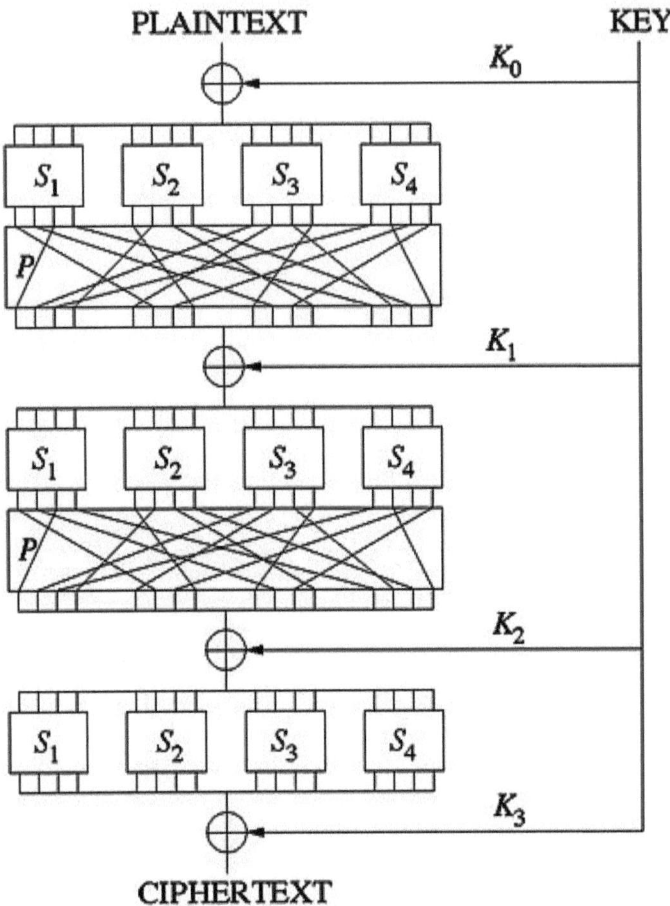

Fig. 1.9 Substitution-permutation network

1.2.4.2   Substitution-Permutation Networks

Substitution-permutation networks (SP networks) represent a class of crypto-graphic techniques widely employed in block ciphers due to their simplicity and effectiveness. These networks operate by alternating rounds of substitution and permutation operations within each block of data, aiming to achieve both confusion and diffusion properties essential for secure encryption. In a basic SP network architecture, as depicted in Fig. 1.9, substitution layers and permutation layers are utilized alternatively across multiple rounds of encryption, each round incorporating a distinct key. This alternating pattern ensures that the plaintext undergoes a series of transformations, rendering it increasingly complex and indistinguishable from the original data. One notable advantage of SP networks is their inherent parallelizability, allowing for efficient implementation on processors with multiple cores. This characteristic makes them particularly well suited for modern computing environments where parallel processing capabilities are prevalent. Notably, SP networks serve as the foundational structure for numerous block ciphers, including the widely acclaimed Advanced Encryption Standard (AES), recognized for its robust security and widespread adoption in various applications [31].

1.2.4.3   DES: Data Encryption Standard

By the early 1970s, the demand for encryption for commercial applications such as banking had become so pressing that it could not be ignored without economic consequences. In 1972, the US National Bureau of Standards (NBS) initiated a request for proposals for a standardized cipher in the USA. The NBS received the most promising candidate in 1974 from a team of cryptographers working at IBM. The algorithm IBM submitted was based on the cipher Lucifer which was developed internally. The algorithm was adopted by the NBS and further developed and published as DES (Data Encryption Standard).

The DES was the first modern cipher that was adopted widely and is the most studied and well-documented cipher. Even until this date, there is no attack better than brute-force attack against the DES cipher. Figure 1.10 shows the DES cipher as a black box. The cipher processes 64-bit data blocks and encrypts them using a 56-bit key. The cipher uses a Feistel network for encryption and is highly efficient in hardware as it was optimized for the 1970s technology.

Figure 1.11 shows the architecture of the DES algorithm. The algorithm utilizes 16 encryption rounds of a Feistel network. Each round of the encryption network uses an F function to introduce the nonlinear element in the cipher. The 64-bit block is split into a right half and a left half. Only the left half is encrypted each round. However, each round, the halves are rotated, ultimately ensuring that all bits are encrypted.

Fig. 1.10 DES as black box

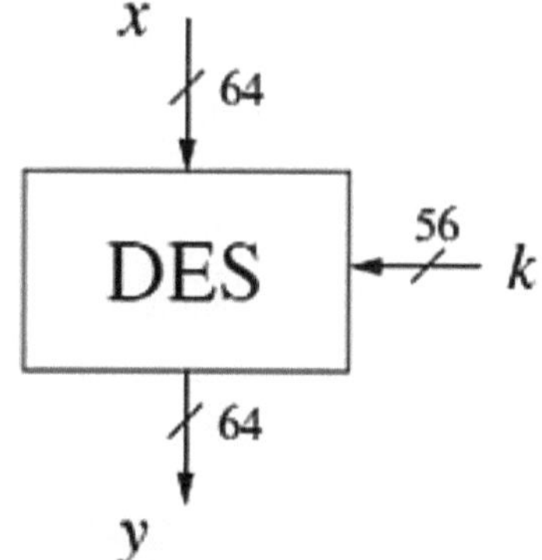

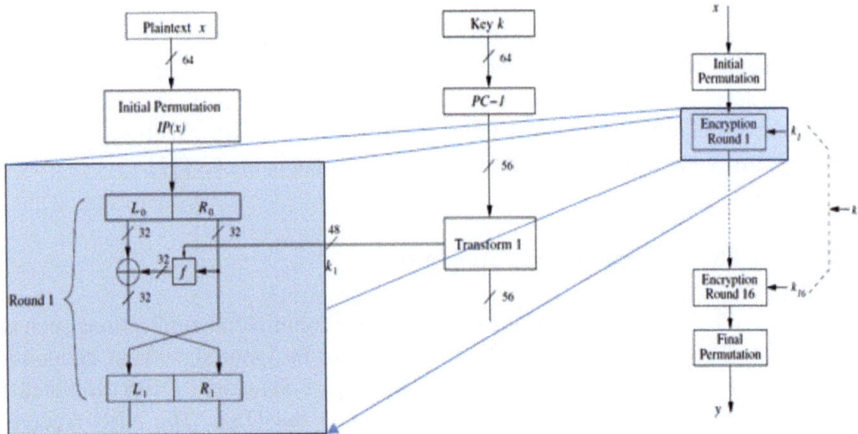

Fig. 1.11 DES architecture

The F function details are shown in Fig. 1.12. The 32-bit half block is expanded into a 48-bit block and XOR-ed with a 48-bit key round. The expansion further enhances the diffusion aspect of the cipher. The 48 bits are then assigned to eight substitution tables, 6 bits at a time. These substitution tables known as the s-boxes (substitution boxes) are the only nonlinear operation in the cipher. The values in these s-boxes are chosen very carefully based on number theory to ensure their resistance against sophisticated attacks like the differential analysis attack [15].

Although the DES algorithm has aged well, it has its issues. DES key space has become too small for current-day computers. A key length of 56 bits is simply not enough to protect against brute-force attacks. To increase the key space, a plain text can be encrypted three times using three different keys, thus increasing the key space to 168 bits. This cipher is simply known as 3DES. However, DES is poorly optimized for software which further makes 3DES too demanding for software implementations. Because of these issues, a demand for a new encryption algorithm has risen, resulting in the development of the AES.

1.2.4.4   AES: Advanced Encryption Standard

As a replacement was needed for the aging DES, in 1997 the National Institute of Standards and Technology (NIST) called for proposals for a new AES. The cipher had to meet the following requirements:

- Block cipher with 128-bit block size
- Three key lengths must be supported: 128, 192, and 256 bits.
- Sufficient security
- Efficiency in software and hardware

Fig. 1.12 The DES F function

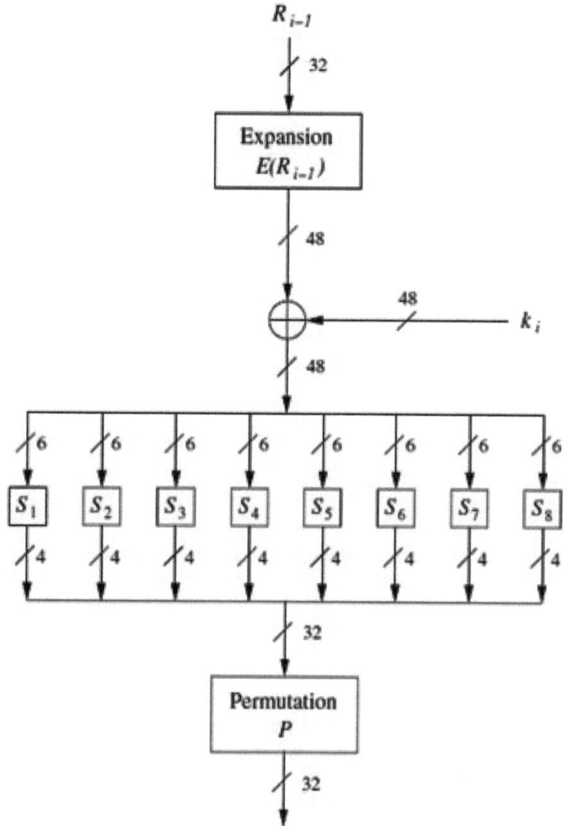

NIST declared the block cipher Rijndael [31] as the new AES and published it as a final standard. The AES is considered the most popular trusted cipher currently employed. It is even adopted by the NSA for encrypting files up to top-secret clearance. AES uses an SP network as opposed to a Feistel network in DES. Similar to DES, the substitution layer includes s-boxes that introduce nonlinearity. The substitution is performed byte-wise instead of bit-wise which makes it software friendly with higher throughput. Figure 1.13 shows a round of AES encryption.

1.2.4.5 PRESENT: Lightweight Block Cipher

PRESENT [17] is a modern cipher that is highly similar to the AES cipher while being capable of operating with an impressive throughput in constraint environments. It is a part of a recent family of ciphers described as "lightweight" due to the reduced cost to implement and reduced power consumption.

Figure 1.14 shows the PRESENT architecture and the s-box it uses. The block length is 64 bits, and two key lengths of 80 and 128 bits are supported. PRESENT

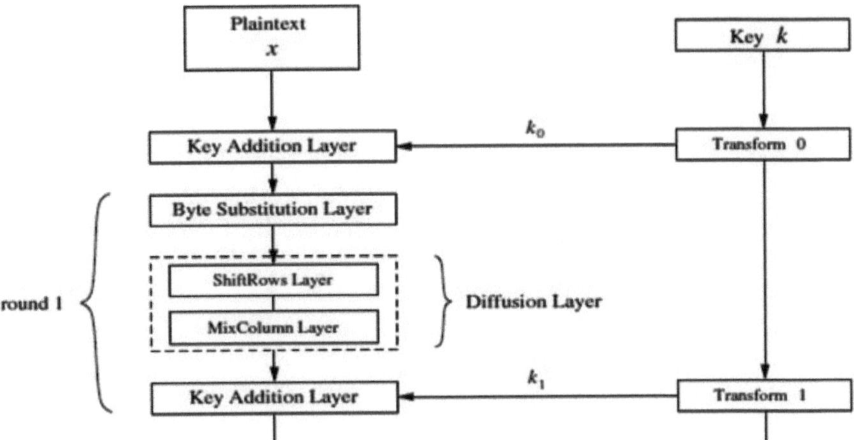

Fig. 1.13 A single AES round architecture

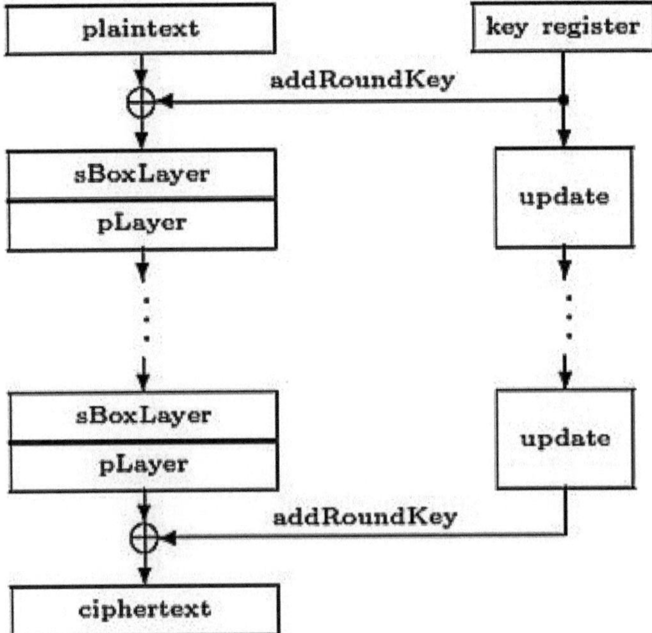

Fig. 1.14 PRESENT cipher architecture

uses smaller 4-bit s-boxes. It also uses fewer, less complex rounds of permutations. PRESENT utilizes an area of 1600 GE for the 80-bit version. However, such area implementation may still exceed the capacity of some constrained devices while offering a cipher that is not as reliable and trustworthy as AES.

1.2.5   Block Encryption Mode

When a block cipher encrypts data one block at a time in a straightforward manner, it would expose the cipher to certain vulnerabilities. If each block of data is encrypted independently by what is known as the Electronic Code Book (ECB) mode, this will make the cipher text vulnerable to different attacks even without breaking a cipher itself.

Consider the message structure shown in Fig. 1.15 used in a bank transaction. Even without breaking the cipher, identifying the blocks of data that correspond to each field will be enough for a substitution attack.

Figure 1.16 shows a different kind of vulnerability. A bitmap image was encrypted using the AES algorithm in the naive ECB mode, and it is clear to the human eye that the encrypted message can still be easily readable. This is the result of the nature of having a deterministic block encryption. To address these vulnerabilities, more complicated block encryption modes that include feedback among the blocks are utilized.

Figure 1.17 shows how a more secure encryption mode, Cipher Block Chaining, would work. The encrypted block from the previous iteration is always XOR-ed with the new data block before the latter is encrypted. This ensures that the data in each block is spread among multiple ones and removes the security vulnerability introduced by encrypting each block independently. The process is easily reversed on the decryption side.

Block #	1	2	3	4	5
	Sending Bank A	Sending Account #	Receiving Bank B	Receiving Account #	Amount $

Fig. 1.15 Fields of an encrypted bank message

Fig. 1.16 Encrypting a figure using a secure cipher

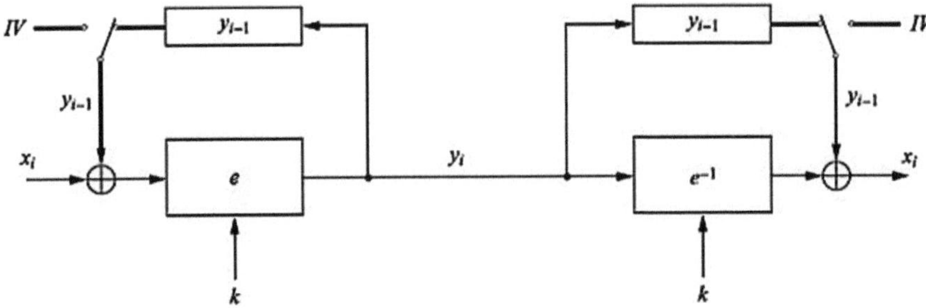

Fig. 1.17 Chained encryption mode

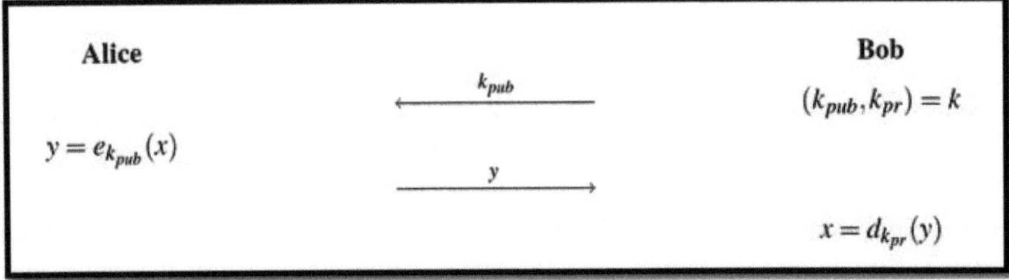

Fig. 1.18 Public-key encryption

1.2.6 Asymmetric Encryption

Managing private keys in a small network can be easily done manually by an operator. However, keys need to be generated and distributed autonomously in large networks. With the possibility of attacks initiating from within the network, careful protocols need to be employed that can thwart such attacks.

While private key encryption ensures that two parties with a common key can establish a secure channel, it offers no solution for securely sharing a secret key. However, public-key encryption protocols that utilize asymmetric ciphers are capable of sharing secret data without the demand for a pre-shared key.

Key encryption is a relatively new branch of cryptology that was discovered in the 1970s. The basic concept behind public encryption is having two keys, one private (used for decryption) and one public (used for encryption). This would allow two parties to share information without anyone eavesdropping. It is usually used to share keys that can be used for private key encryption. Figure 1.18 illustrates the concept.

The encryption function used in public-key encryption should be a one-way function. A function $f()$ is a one-way function if:

1. $y = f(x)$ is computationally easy.
2. $x = f^{-1}(y)$ is computationally infeasible.

While there are several mathematical functions that can satisfy these two criteria, so far, there have been three popular mathematical problems that have been used in asymmetric ciphers:

1. Integer-Factorization Schemes: Several public-key schemes are based on the fact that it is difficult to factor large integers, while multiplying large integers is very easy. The most prominent representative of this algorithm family is RSA [127].
2. Discrete Logarithm Schemes: There are several algorithms that are based on what is known as the discrete logarithm problem in finite fields. The most prominent examples include the Diffie–Hellman key exchange, Elgamal encryption [41], or the Digital Signature Algorithm (DSA) [47].
3. Elliptic Curve (EC) Schemes: A generalization of the discrete logarithm algorithm is elliptic curve public-key schemes. The most popular examples include Elliptic Curve Diffie–Hellman (ECDH) key exchange [8] and the Elliptic Curve Digital Signature Algorithm (ECDSA) [47].

In addition to making secure key establishment possible with no pre-shared information, asymmetric encryption provides the means to establish assurance and trust in many applications. We list its more important uses:

- Key Establishment: There are protocols for establishing secret keys over an insecure channel. Examples of such protocols include the Diffie–Hellman Key Exchange (DHKE) or RSA key transport protocols.
- Non-repudiation: Providing non-repudiation and message integrity can be realized with digital signature algorithms, e.g., RSA, DSA, or ECDSA.
- Identification: Identify entities using challenge-and-response protocols together with digital signatures, e.g., in applications such as smart cards for banking or for mobile phones.
- Encryption: Encrypt messages using algorithms such as RSA [127] or Elgamal [41].

1.2.7   Side-Channel Attacks

Side-channel attacks deserve a special mention as they can extract the private key without breaking the cipher or compromising its security. This is done by exploiting information about the private key, which is leaked through the device's physical channels. An example would be power trace timing analysis. For this attack to work, the attacker must have direct physical access to the device.

Figure 1.19 shows a power trace of the processor performing RSA decryption. While the trace shows a naive implementation, it demonstrates how such implementations can leak information about the secret key.

The algorithm uses a square (S) cycle for a 0 bit in a key and a square and multiply (SM) cycle for a 1 bit in a key. Since the SM cycle takes around twice the time to execute, we can easily identify the alternating S cycles that correspond to 0

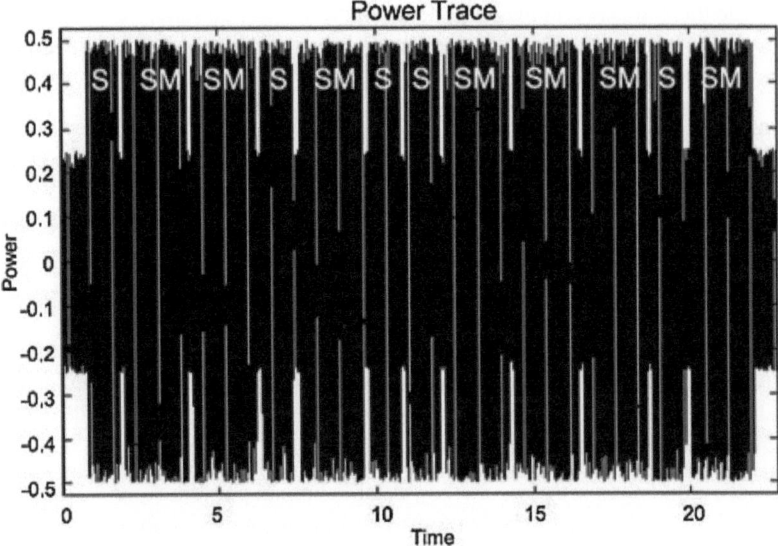

Fig. 1.19 Power trace can expose key in a naive implementation

and the *SM* cycles that correspond to 1 in the key. Even without breaking the cipher, the key was obtained. This shows the power of side-channel attacks. Hardware and software implementations of ciphers need to address such attacks when it is feasible for adversaries to have access to side-channel attacks.

1.2.8 Hash Functions

Hash functions serve as critical components in modern cryptography, offering a distinct approach to data transformation known for its one-way nature. Unlike encryption algorithms, which typically involve reversible transformations between plaintext and ciphertext using secret keys, hash functions operate unidirectionally, generating fixed-size hash values from variable-length input data. This means that even the slightest alteration in the plaintext results in a completely different hash value, making hash functions invaluable for data integrity verification and authentication purposes. As illustrated in Fig. 1.20, a small change in the input text leads to a drastically different output hash, highlighting the deterministic yet irreversible nature of hash functions. Since hash functions do not rely on secret keys for their operation, they find extensive use in scenarios where verification and authentication are paramount, such as digital signatures and password hashing. By employing hash functions alongside encryption, message authentication can be ensured, safeguarding against unauthorized alterations to the message content by potential adversaries. Prominent examples of hashing algorithms widely utilized

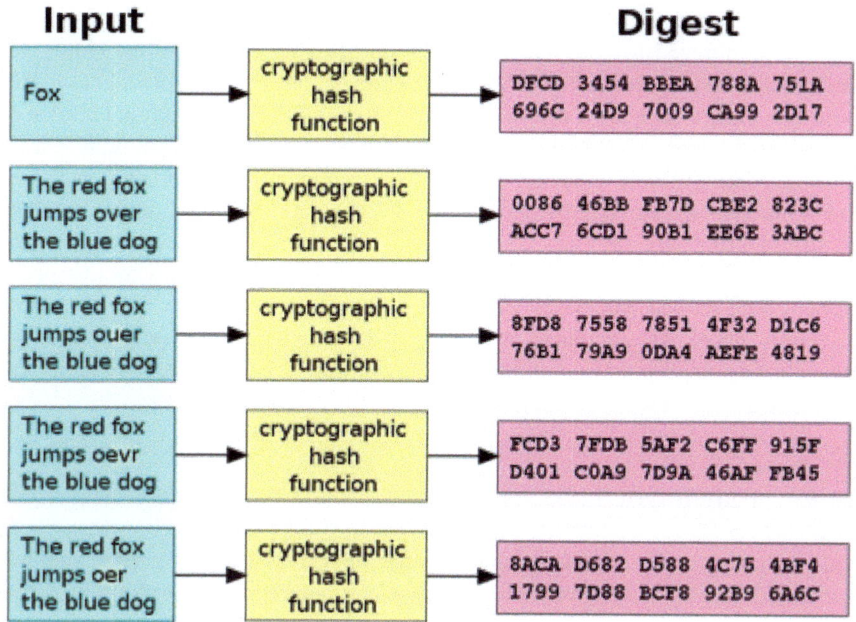

Fig. 1.20 Hashing is used to detect any modification to original text

in cryptographic applications include SHA-1 and SHA-2, renowned for their robustness and widespread adoption across various security protocols [39, 40].

1.2.9 Security Classes

Security protocols encompass a broad spectrum of computational complexities, reflecting the diverse requirements and constraints imposed by different application scenarios. These protocols can be broadly categorized into four distinct classes based on the computational complexity of the algorithms they employ, as delineated by Sasi [30]. First, full-fledged security protocols represent the most comprehensive category, characterized by the utilization of sophisticated cryptographic techniques and algorithms. These protocols often involve complex computations and are typically employed in scenarios where stringent security requirements necessitate robust cryptographic solutions, such as in enterprise-grade communication networks and secure data transmission over the Internet.

Moving toward the other end of the complexity spectrum, we encounter lightweight security protocols, which offer a balanced trade-off between computational efficiency and security robustness. These protocols are designed to address the resource-constrained nature of certain devices and systems, such as IoT devices and wireless sensor networks, where computational overhead must be minimized to conserve energy and processing resources. By employing streamlined cryptographic algorithms and optimized protocols, lightweight security protocols ensure adequate protection while mitigating the computational burden on constrained devices. Furthermore, ultralightweight security protocols represent the most resource-efficient category, tailored for deployment in extremely constrained environments where computational capabilities are severely limited. These protocols prioritize minimal computational overhead, often sacrificing certain security features to accommodate the stringent constraints imposed by low-power devices with limited processing capabilities. Overall, the categorization of security protocols based on computational complexity provides a valuable framework for selecting appropriate solutions that align with the specific requirements and constraints of diverse application scenarios.

1.2.9.1 Full-Fledged Class

Full-fledged security protocols represent the pinnacle of cryptographic sophistication, leveraging a comprehensive array of cryptographic algorithms to fortify the security and privacy of communication channels. Within this class, a diverse arsenal of cryptographic primitives is deployed, encompassing both symmetric and asymmetric cryptographic techniques, along with robust hashing functions. Symmetric cryptographic ciphers, such as AES, serve as the cornerstone for ensuring confidentiality by encrypting data using a shared secret key, while public-key encryption protocols like RSA facilitate secure communication channels by enabling key exchange and digital signatures. Additionally, hashing functions like SHA-256 play a vital role in data integrity verification, providing assurances that transmitted data remains unaltered during transit. By harnessing the combined strength of these cryptographic tools, full-fledged security protocols can deliver robust protection against a wide spectrum of security threats, ensuring the confidentiality, integrity, and authenticity of sensitive information in communication networks.

Furthermore, the efficacy of full-fledged security protocols hinges on their proper implementation and configuration, which entails adherence to established cryptographic best practices and protocols. Proper key management practices, including secure key generation, distribution, and storage, are essential for maintaining the confidentiality of encrypted data. Moreover, stringent authentication mechanisms, such as digital certificates and multifactor authentication, bolster the resilience of these protocols against unauthorized access and impersonation attacks. By diligently following security protocols and employing rigorous cryptographic measures, organizations can harness the full potential of full-fledged security protocols to establish a robust security posture that safeguards sensitive communication channels against evolving cyber threats.

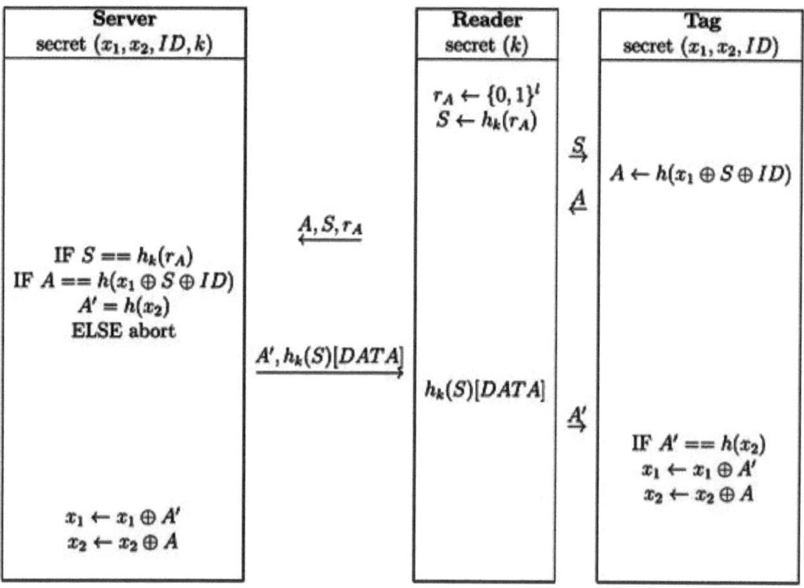

Fig. 1.21 Yang et al.'s hash-based protocol

1.2.9.2 Simple Class

Protocols in the "simple" class would avoid the use of encryption and rely instead on Random Number Generators and hash functions to secure their communication. While the hash functions have a high computational complexity, by avoiding the use of symmetric and asymmetric ciphers, the overall complexity of such protocols would be much lower than the full-fledged ones.

Figure 1.21 shows Yang's [164] hash-based mutual authentication protocol. Secret values are hashed before transmission. To protect against replay attacks, secret values are updated at the end of a successful authentication. This protocol is vulnerable to desynchronization attacks as the last sent message can be interrupted, and only one of the parties would update its secret values, causing the other party to be out of sync. Secure hashing algorithm has high implementation requirements that are not suitable for most constrained devices.

1.2.9.3 Lightweight Class

Lightweight protocols represent a pragmatic approach to security, particularly in resource-constrained environments where computational complexity is a primary concern. Unlike their full-fledged counterparts, lightweight protocols eschew the use of complex cryptographic primitives in favor of simpler alternatives, such as Random Number Generators (RNGs) and Cyclic Redundancy Checks (CRCs). By

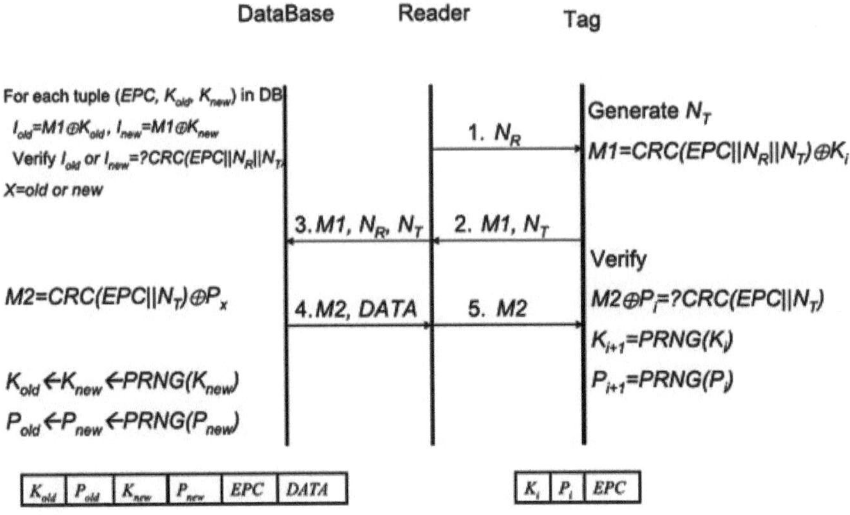

Fig. 1.22 Chien and Chen's lightweight protocol

leveraging RNGs, lightweight protocols introduce randomness into the communication process, enhancing unpredictability and resistance against certain types of attacks. Additionally, CRCs serve as a lightweight alternative to cryptographic hashing functions, enabling data integrity verification without the computational overhead associated with traditional cryptographic hashing algorithms.

Chien and Chen's protocol, depicted in Fig. 1.22, exemplifies the principles of lightweight security protocols by incorporating CRCs and PRNGs into the communication framework. While offering computational efficiency, the protocol's reliance on CRCs exposes it to vulnerabilities, as CRC was originally designed for error detection rather than cryptographic security. As a result, CRCs lack the robustness and resistance against manipulation required for secure communication in adversarial environments. Despite their inherent limitations, lightweight protocols like Chien and Chen's protocol remain relevant in scenarios where computational resources are scarce, albeit with the caveat that their security posture may be compromised in the face of determined attackers.

1.2.9.4 Ultralightweight Class

Ultralightweight protocols represent the most minimalist approach to security, relying solely on basic bitwise operations such as XOR, AND, OR, and addition modulus 2. These protocols prioritize computational simplicity and efficiency above all else, making them suitable for extremely resource-constrained environments where even modest cryptographic primitives are impractical. One such example is the SASI protocol, illustrated in Fig. 1.23, which demonstrates the application

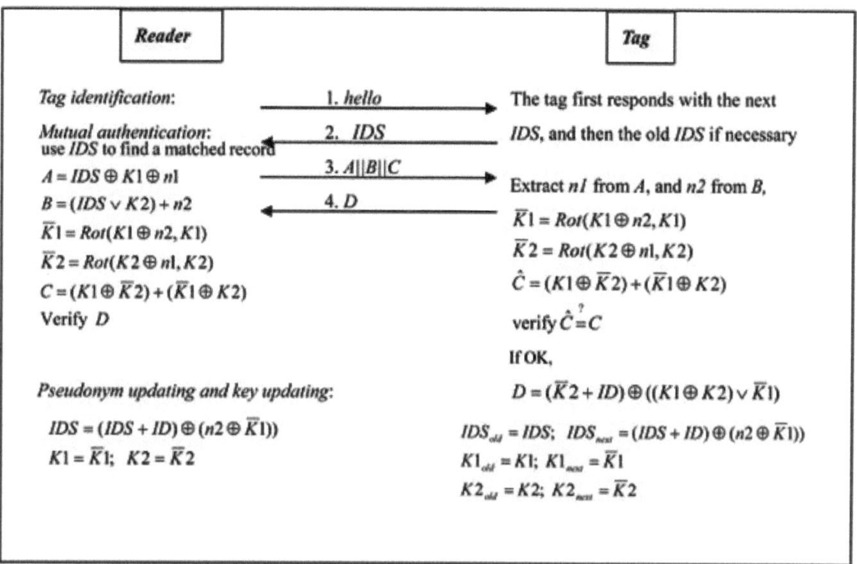

Fig. 1.23 Protocol steps of SASI

of ultralightweight techniques in securing communication channels. By employing a combination of addition and rotation operations, the SASI protocol aims to obscure secret values within the data stream. However, the linear nature of these operations renders the protocol vulnerable to determined attackers with access to sufficient computational resources. Despite their inherent limitations in providing robust security guarantees, ultralightweight protocols like SASI remain relevant in scenarios where computational overhead must be minimized at all costs, albeit at the expense of compromising certain aspects of security.

Chapter 2
Exploring Classic Security Attacks and Hardware Intricacies

The digital age has ushered in an era where electronic devices are integral to nearly every aspect of our daily lives, from personal communication to critical infrastructure. This ubiquity has also brought about significant security challenges, especially with the exponential growth of Internet of Things (IoT) devices and their interconnected networks. The increasing complexity and widespread adoption of these technologies have introduced new vulnerabilities that malicious actors can exploit. Therefore, it is crucial to understand and prepare for these potential threats by designing robust security measures and countermeasures [2, 3, 78, 141]. This chapter aims to organize and elucidate the complex field of hardware security by exploring classic security attacks and the intricate workings of hardware systems that can both mitigate and contribute to these security challenges, as shown in Table 2.1.

2.1 Classic Security Attacks

Electronic devices are a significant part of our daily lives, and we will continue to depend on these devices more and more every day. The exponential growth of the IoT devices and their ubiquitous connections are creating more security vulnerabilities. Now, it is high time to be better prepared to tackle these attacks and design secure methods to protect and provide countermeasures for these attacks, to make these complex hardware security fields a bit more organized.

2.1.1 Hardware Trojans

Hardware Trojans mean inserting malicious modifications to the circuit. Internal complexity, scaling down of IC to the physical limit, and unspecified space available

© The Author(s), under exclusive license to Springer Nature Switzerland AG 2025

K. Khalil et al., *Lightweight Hardware Security and Physically Unclonable Functions*, https://doi.org/10.1007/978-3-031-76328-1_2

Table 2.1 Overview: exploring classic security attacks and hardware intricacies

Section	Description
Classic Security Attacks	Overview of security vulnerabilities in electronic devices, emphasizing the need for robust countermeasures
Hardware Trojans	Insertion of malicious modifications into circuits, detection challenges, and threat models
IP Piracy and IC Overbuilding	Security threats arising from outsourcing IPs, including IP cloning, imitation, and Trojan insertion
Side-Channel Analysis	Exploitation of information leakage via physical modalities to extract secrets from target systems
Counterfeiting	Unauthorized imitation or forgery of semiconductor components, impacting system performance and reliability
Classic Hardware Attacks	Various methods for compromising hardware systems, including packet header attacks, tampering, and MitM attacks
Attack on Deep Neural Networks	Hardware-based attacks on neural networks, including training-time Trojan attacks and hardware intrinsic attacks on CNN layers
Software Attack to the Hardware	Execution of traditionally software-based attacks through hardware, such as ransomware and timing-based side-channel attacks

in the IC are some causes why hardware Trojan detection is difficult. There are two types of threat models; the first one is manipulating the lithographic masks, which is when the foundry inserts a Trojan into the design in the form of addition, deletion, or modification of gates [75, 150].

The second type is when a malicious insider does not provide information about hardware Trojan to the validation team of malicious circuitry. However, hardware Trojan can be inserted (modified) by a third-party IP design house. There are two types of Trojan detection techniques: One is called invasive (and semi-invasive), which requires costly, precision measurement equipment, and components become unusable afterward. On the other hand, noninvasive detection methods depend on external parametric and functional IC testing, such as input and output pattern, delay, and dynamic leakage. Self-monitoring [145] and static verification [66] are two protections against rogue 3PIP and insider attacks.

2.1.2 IP Piracy and IC Overbuilding

Modern SoC and IC designs usually need different forms of IPs, for example, Register Transfer-Level (RTL) design (soft IP), gate-level netlist (firm IP), and physical layout (hard IP) [68]. The companies outsourced IPs from trustworthy offshore design houses or foundries for SoC integration or IC fabrication to reduce design complexity, time-to-market pressure, and manufacturing cost. This globalization of outsourcing IPs can lead to various IP security threats, such as IP cloning, IP imitation, and inserting Trojans inside the IPs/ICs. This type of hardware attack happens if an outsider has excess to an IP or an IC and illegally claims ownership or steals the IP to overbuild and sell them [25, 130]. Attackers from the integration house or foundry can pirate 3PIP and use more than the amount of 3PIP instances allowed by the license, extracting information from the design layout, the IC design, and/or overbuilding it accordingly. Moreover, the fraudulent designs not only harm the reputation and earnings of the legitimate vendor but also seriously negatively affect important systems' performance and reliability because they contain harmful or backdoor logic that might leak private data or help override them [88].

2.1.3 Side-Channel Analysis (SCA)

Side channels are instances of inadvertent information leakage where an attacker attempts to obtain data from a target computing system by exploiting technical flaws. Side-channel attacks attempt to extract secrets through a physical modality, for example, power consumption [85], electromagnetic (EM) emanations [128], photonic emissions [138], acoustic noise of the system [56], etc. An attacker can use a different method to measure the switching power traces of an electronic device during its operation, and after that, the attacker can use logical formulation to extract secret keys or secret data. Simple Power Analysis (SPA) and Differential Power Analysis (DPA) are methods for obtaining confidential information from a power SCA. Another technique for analyzing power channels is called Correlation Power Analysis (CPA) [21], which uses statistical models to calculate the correlation between the secret and the device's power usage when the secret is used for computing. Additionally, EM radiation during the execution of security-critical activities results in hardware vulnerability due to data leakage. EM radiation during the operation of security-critical processes also causes hardware vulnerability through leaking information [123], and many of these EM traces are helpful to extract secret keys. Leakage reduction, noise injection, key updates, side-channel resistance Physical Unclonable Function (PUF), and Secure Scan Chains are a few examples of protection against side-channel attacks.

2.1.4 Counterfeiting

Counterfeiting is frequently carried out by numerous companies involved in the semiconductor supply chain, such as new product suppliers or secondary (recycled) IC vendors. Selling used chip as new IC also falls into this category. An unauthorized imitation or forgery of the original semiconductor component is known as a counterfeit semiconductor component. One of the many organizations in the semiconductor supply chain, such as suppliers of new products or secondary (recycled) IC suppliers, frequently engages in counterfeiting. The ability to discern between fake and genuine ICs has recently improved because of technological advancements in 3D packaging.

The manufacturers of authentic components lose money as a result of counterfeiting ICs in the market. The performance and dependability of the entire system are negatively impacted by the mediocre performance of Counterfeit goods, usually in lower quality or older versions of the original product. The authentic provider's reputation is also harmed. Such counterfeit goods could potentially hamper the performance of critical applications in healthcare, banks, cars, planes, and weapons control systems and cause lots of financial loss. It was calculated that US corporations suffered financial losses of between \$225 and \$600 billion in 2017 [16] due to design IP infringement. Some defense techniques for hardware attacks are hardware metering and auditing (tracking of manufactured integrated circuits after fabrication), IC Fingerprints or PUFs [37, 110], etc. Negative temperature bias instability (NBTI), hot carrier injection, and electromagnetic migration are some parameters that influence the lifetime of an IC, which can be estimated by the related sensor.

2.1.5 Classic Hardware Attack

Classic hardware attacks represent a category of sophisticated techniques employed by adversaries to compromise the security of hardware systems. These attacks exploit vulnerabilities at the physical level of hardware components, bypassing traditional software-based security measures and directly targeting the underlying hardware implementation. Classic hardware attacks encompass a diverse range of methodologies, including invasive techniques such as microprobing and side-channel analysis, as well as noninvasive approaches like fault injection and electromagnetic analysis. By exploiting inherent weaknesses in the design or implementation of hardware systems, adversaries can gain unauthorized access, extract sensitive information, or tamper with system functionality. Understanding classic hardware attacks is essential for developing robust countermeasures and ensuring the integrity and confidentiality of hardware-based systems in the face of evolving security threats.

2.1.5.1 Packet Header Attack by Hardware Trojan

Kulkarni et al. [89] model an HT that can alter the destination address field of specific NoC packets and is mounted on the input buffers of NoC routers. Then, it was simulated in a TCMP, whose effect led to packet drop upon reaching a wrong destination tile. This paper also finds a suitable area within the NoC router where an HT can install itself and change packet content without interfering with its fundamental functionality. The proposed HT can severely affect the miss penalty of L1 caches and even stop an application by blocking an instruction. However, this only applies to situations where the cores perform low injection rate programs.

2.1.5.2 Tampering Attack on AES Cores with Hardware Trojans

The study by Jain et al. [73] describes a unique tampering technique using a sequential hardware Trojan to modify and obtain the secret key from the AES core. The proposed method depends on when an untrusted foundry inserts a hardware Trojan in the netlist. The proposed Trojan activation can modify the input data of a particular round by preventing key-dependent computation of the previous round. Then, the proposed sequential hardware Trojan helps the attacker to collect this modified data at the input of an internal round. In the proposed method, Trojan lets an attacker calculate the last round subkey using the observed ciphertext. An algorithm to recover the original key from the subkey from the previous round was provided in this paper.

Since the attacker only knows the trigger condition generated from the input plaintext and applies it repeatedly, it is very challenging to identify such Trojans using logic testing. Overhead is caused by the payload because it needs 128 OR Gates.

2.1.5.3 Hardware Trojans Attack on Analog/Mixed-Signal ICs

Designing HTs for analog ICs is highly difficult because it is hard to meet all the requirements for a successful HT. First, as analog signal pathways are usually susceptible, it is challenging to design stealthy HT because doing so could result in nonnegligible performance deterioration. Second, because analog designs often have few components or can be easily separated into subblocks or stages with a few components apiece, it is challenging to create tiny footprint HTs that will escape optical reverse engineering. Third, as the researchers can extract multiple information-rich measurements from any analog IC, testing will likely be able to distinguish HT-infected and HT-free instances. Despite all the difficulties mentioned above, designing an analog Trojan is possible.

The paper by Elshamy et al. [43] proposed a possible HT attack scenario in which the Trojan does not reside inside the analog circuit of the victim. However, it is located on the independent digital circuit on the same die where it was triggered. Its payload is only applied to the analog circuit after being transmitted via the common test infrastructure and the analog circuit's test interface. This HT attack uses the dense digital circuit to conceal its footprint successfully, making it impossible to detect in the analog domain.

2.1.5.4 Man-in-the-Middle Attack on PCIe Bus for Smart Data Replay

The architecture of a Stratix 5 FPGA-based platform is proposed by Khelif et al. [82] to conduct a hardware Man-in-the-Middle (MitM) attack on a computer's PCIe 3.0 bus and emulate the behavior of an SoC. This MitM can log, edit, and copy data into a shadow memory so that it can be replayed to the host. The Man-in-the-Middle attack, expressed as MitM, consists of inserting a third device into an end-to-end communication. Between the two peripherals, this device behaves as an invisible rooter. By recording and interfering with the transmitted data in real time, the attacker in this position has the ability to undermine the communication's confidentiality and integrity. Hardware MitM is a special attack that provides direct access to data and offers various attack vector possibilities, including data replay, traffic analysis, and fault injection. It is not only one of the least exploited attacks in the state of the art but also one of the most promising attacks. All the information kept in memory, including the counter for password attempts, is encrypted, but a hardware MitM attack may still damage it.

2.1.5.5 Hardware Trojan Attacks on Processor

Conditionally triggered hardware Trojan attacks on processor cores based on time information as the trigger conditions. Research by Kuo et al. [90] presents two types of hardware Trojan designs depending on the time information as the trigger conditions, one using real-world time and the other using the time passed since power on. A programmable scheme was also designed in this study to set a trigger condition by controlling the scan chain of a system. This work also showed how to insert these Trojan designs in an open-source processor core (RISC-V processor). This type of Trojans typically comprises two parts: the trigger and the payload. The trigger part monitors the internal or external signals of the host circuit. When some specific, rare condition is satisfied, the payload part starts to attack the host design. The trigger part can be implemented using combinational, sequential, or both logics. The attack has a low area and power overhead as they have used existing circuit components along with a counter to design the Trojan.

2.1.5.6 Hardware-Oriented Algebraic Fault Attack

Gay et al. [55] proposed a framework that can construct fault attacks automatically on hardware-oriented structural cipher descriptions, which can quickly evaluate any cipher implementations, including any optimizations. However, the output of the framework is a set of algebraic equations that are then fed to a SAT solver. It takes two primary inputs, namely the circuit model and the fault descriptions. It takes two primary inputs: the circuit model and the fault description. The fault description mainly consists of three parts: (1) the fault round (e.g., in which round or at which operation the fault is injected), (2) the fault location (e.g., the byte or nibble affected by the fault), and (3) the number of faults. The proposed framework is the first Algebraic Fault Attack (AFA) tool to examine full-scale AES, LED-64, and PRESENT with only hardware-oriented structural cipher descriptions.

2.1.6 Attack on Deep Neural Network

Attacks on deep neural networks (DNNs) represent a significant concern in the realm of machine learning security, posing threats to the integrity and reliability of artificial intelligence systems. As DNNs continue to be deployed in various applications, ranging from image recognition to autonomous vehicles, they become lucrative targets for malicious actors seeking to manipulate their behavior or exploit vulnerabilities. Attack methodologies targeting DNNs are diverse, encompassing adversarial examples, model inversion, data poisoning, and backdoor attacks, among others. These attacks aim to deceive or undermine the functionality of DNNs by exploiting their inherent vulnerabilities, such as sensitivity to perturbations in input data or reliance on inadequately trained models. Understanding the nature of attacks on DNNs is crucial for developing robust defense mechanisms and ensuring the trustworthiness of AI-driven systems in critical domains.

2.1.6.1 Training-Time Trojans Attacks (DNN Trojan Attacks)

Model Trojaning attacks, in which adversaries introduce malicious backdoors into ML models, are one of the most severe security vulnerabilities in DNNs. DNN Trojan attacks can cause a victim model to purposefully misclassify a subset of inputs via a Trojan trigger, with little to no impact on the inference of actual inputs. Previous training-time attack literature can typically be divided into two groups: (i) Data poisoning attack [28] is when victim models are trained using maliciously constructed poisoned data samples, and (ii) a retraining-based Trojan attack is when adversaries can arbitrarily alter model parameters through local training [100]. One recently reported hardware-based fault attack is rowhammer [84], which demonstrated practical concerns about the integrity of computing systems. This attack is more severe because fault attack vectors allow internal tampering of a target system even when the data and software stack are configured correctly and controlled by legitimate users. The recent study also shows inference time adversarial attacks, which inject bit-flips into DNN models stored in memory to compromise model classification behaviors [22]. Moreover, a recent work by Cai et al. [23] provides hardware-induced memory faults as a means of training-time Trojan attacks. During machine learning training, the proposed attack methodology perturbs the feature maps of a victim model via bit-flips and finds the Trojan trigger pattern during inference time. This framework provides a high attack success rate on average of 99.98% on misclassification.

2.1.6.2 Hardware Trojan Attack on Activation Parameters of FPGA-Based DNN Accelerators

The work by Mukherjee et al. [118] proposes a noble hardware Trojan-based attack on the internal activation parameters of two trained DNN hardware accelerators. The trained DNN hardware accelerators are intended for object identification and image classification and are targeted toward the FPGA platform. When successfully triggered, the glitch-generating circuitry in the proposed HT modifies the activation parameters of the DNN in such an accelerator. The FPGA platform is the intended target for the trained DNN hardware accelerators, which are designed for object detection and image classification.

2.1.6.3 Hardware Intrinsic Attack (HIA) on Convolutional Neural Network (CNN)

Odetola et al. [121] presented noble HIA. The proposed HIA is inserted inside mathematical operations of specific CNN layers, propagating erroneous operations in all the succeeding CNN layers and resulting in misclassification. The attack propagates incorrect feature mappings by rearranging the weight matrices of convolutional layers. This work also shows five alternative attack scenarios that are developed for each CNN layer and assessed in terms of the rate of triggering and overhead resources compared to the initial implementation. Additional latency is minimal in all attack scenarios (0:61%), and there is also a minimal increase in DSP, LUT, and FF (2:36% in each case). Only very slight variations in BRAM are evident in three of the five tested scenarios. The parameters of the model are not altered to carry out this attack. Attacks are entirely random and nonperiodic, making them hard to identify.

2.1.6.4 Multichannel Fusion Attack Framework Based on DNNs Against Side-Channel Protected Hardware

A multichannel fusion framework based on DNNs is proposed by Hettwer et al. [65]. The proposed method extracts location-dependent leakages that capture the activity of specific parts of the hardware design using different decoupling capacitors on a modern system-on-chip (SoC) using 16-nm fabrication technology of the power supply. A multi-input DNN is then used to analyze the collected data. This network has been taught to aggregate share-related data in order to forecast the results of processed masked data. This method works with more realistic adversary assumptions as no mask values need to be known for a successful attack.

2.1.7 Software Attacks to the Hardware

Software attacks on hardware represent a formidable threat landscape where vulnerabilities in software are exploited to compromise the underlying hardware components. These attacks target various layers of the hardware stack, ranging from firmware and device drivers to operating system kernels and applications, leveraging software vulnerabilities to gain unauthorized access, manipulate system behavior, or extract sensitive information. Examples of software attacks on hardware include buffer overflows, privilege escalation exploits, rootkits, and side-channel attacks, each exploiting different aspects of software design and implementation to compromise hardware security. As software continues to play a crucial role in controlling and managing hardware functionality, understanding and mitigating software-based threats are paramount for ensuring the integrity and resilience of modern computing systems.

2.1.7.1 EthClipper Hardware Wallet Attack

Ivanov et al. [72] proposed an attack named EthClipper, an attack that goes after owners of hardware wallets on the Ethereum platform. EthClipper is a Hardware Wallet Clipboard Meddling Attack with Address Verification Evasion. To choose the address with the highest visual similarity to the original one, the EthClipper malware searches a distributed database of pre-mined accounts. EthClipper adds a social engineering component to the existing clipboard substitution technique. A distributed system named ClipperCloud is used by the hacker in EthClipper to mine and store billions of Ethereum accounts. When the malware discovers an Ethereum address in the clipboard, it instructs ClipperCloud to find the mined account among the accounts generated with the greatest visual resemblance to the address in the clipboard. Because the address on the screen and the expected address appear to be visually similar, the victim's confirmation bias may be triggered, resulting in the victim's approval of the fraudulent transaction. To carry out this attack, only a limited budget is required, and it can cause malicious transactions.

2.1.8 Known Software Attacks on Hardware

Known software attacks by hardware refer to common attack approaches in which hardware components are manipulated or exploited through vulnerabilities in software systems. These attacks exploit weaknesses in software layers to compromise the functionality or security of hardware devices. Examples of known software attacks on hardware include malware that targets firmware or device drivers, vulnerabilities in operating systems that allow unauthorized access to hardware resources, and attacks that leverage software bugs to trigger unintended behavior in hardware components. Understanding these known software attacks is crucial for developing effective countermeasures to protect hardware systems from exploitation and ensure their continued security and reliability in the face of evolving threats.

2.1.8.1 Hardware Ransomware Attack

Almeida et al. [6] proposed a ransomware attack which was previously existed in the software domain only. Ransomware is a category of malicious software (sometimes known as malware) intended to encrypt or restrict user access to data [26].

Typically, a ransomware attack is triggered by the user or a timing event, and all user data is encrypted. A specialized hardware Trojan attack called hardware ransomware attack was proposed in this study by combining cryptographic hardware and a key generation scheme. This study examined the viability of hardware ransomware attacks and offered comprehensive instructions for mapping the design to multiple hardware platforms, including FPGA and ASIC, two along with a silicon implementation in 65 nm CMOS. It was claimed that ransomware attacks are possible on the hardware.

2.1.8.2 Timing-Based Side-Channel Attack on Nonpersistent Network-on-Chip (NoC) Hardware

Ali et al. [5] presented a sophisticated timing-based side-channel attack targeting nonpersistent Network-on-Chip (NoC) hardware, highlighting its practical application on a real multicore machine. Their innovative attack setup, named ConNOC, capitalizes on interference within the NoC hardware to reduce noise and increase predictability, thus enhancing the effectiveness of the attack. By meticulously orchestrating the interference, ConNOC achieves an 11-fold improvement over baseline methods when tested on an actual CPU. Remarkably, ConNOC demonstrates 100% accuracy in both covert communication and information leakage scenarios, even when employing five replays to transmit information. This significant advancement underscores the potential vulnerabilities in NoC hardware and emphasizes the need for robust security measures to mitigate such side-channel attacks. The study by Ali et al. not only showcases the feasibility of these attacks but also serves as a crucial call to action for the development of more secure NoC architectures.

Acknowledgments This chapter was contributed by Sonia Akter. The material presented in this chapter is based on the survey paper: "Sonia Akter, Kasem Khalil, and Magdy Bayoumi. A survey on hardware security: Current trends and challenges. *IEEE Access*, 2023."

Chapter 3
Hardware Security Challenges

Hardware security research methods face a myriad of challenges that need to be meticulously addressed to develop more advanced and effective solutions. As technology continues to evolve, so do the threats and vulnerabilities associated with hardware systems. This chapter delves into the most common challenges in the field of hardware security, emphasizing their significance and the urgent need for robust and innovative solutions.

One of the primary challenges arises from the globalization of the integrated circuit (IC) supply chain, which introduces diverse vulnerabilities due to the outsourcing of IP and fabrication processes. As companies seek to reduce production costs and meet time-to-market demands, they increasingly rely on third-party vendors, amplifying the risks of tampering and intellectual property theft. Additionally, the rapid expansion of Internet of Things (IoT) devices has brought about concerns regarding the security of energy-limited and lightweight devices. These devices often operate under strict energy constraints, making it difficult to implement traditional security measures without compromising performance and efficiency.

Data privacy and confidentiality are also critical issues, particularly in an era where data breaches can have devastating consequences. Ensuring that sensitive information, such as cryptographic keys, remains secure is a formidable task, especially when considering the potential for side-channel attacks that exploit physical leakages. Furthermore, the need for flexible and reconfigurable security mechanisms adds another layer of complexity. While Application-Specific Integrated Circuits (ASICs) offer optimal performance for specific tasks, they lack the flexibility required to adapt to new threats, unlike Field-Programmable Gate Arrays (FPGAs), which, despite their flexibility, present additional security challenges.

The advent of advanced chip packaging technologies aimed at achieving high performance and cost efficiency has also introduced new security risks. The intricate nature of modern packaging methods makes it difficult to detect tampering and ensure the integrity of ICs. Moreover, the backside vulnerability of silicon chips,

35

K. Khalil et al., *Lightweight Hardware Security and Physically Unclonable Functions*, https://doi.org/10.1007/978-3-031-76328-1_3

particularly in flip-chip packaging, exposes critical data to potential attacks, further complicating the security landscape. In this chapter, the most common challenges are described as follows:

3.1   Outsourcing IP/Integrated Circuit

The fabrication process in the integrated circuit (IC) industry has undergone significant transformations due to the globalization of the semiconductor sector over the past decade [44]. This globalization has led to a more globally diversified IC supply chain, driven by the need to meet time-to-market demands and manage the steadily increasing complexity of hardware designs [142]. To remain competitive and reduce production costs, companies are increasingly outsourcing intellectual property (IP) from third-party vendors. This practice allows firms to leverage specialized expertise and advanced technologies without incurring the high costs associated with in-house development. However, this diversification of the hardware supply chain introduces a broader scope of potential vulnerabilities. Each outsourced component or IP block presents a potential entry point for malicious actors, who could exploit these elements to introduce security risks such as hardware Trojans, counterfeits, or IP theft. As a result, while outsourcing can provide significant economic and logistical advantages, it also necessitates stringent security measures and comprehensive vetting processes to ensure the integrity and security of the entire IC supply chain. This complex landscape underscores the need for robust supply chain management practices and the development of advanced security protocols to safeguard against the diverse threats emerging from a globally distributed fabrication ecosystem.

3.2   Security for Energy Limited Devices/Lightweight Devices

The rapid proliferation of IoT devices has heightened concerns about ensuring reliable and secure communication among these devices. Given that many IoT devices operate under stringent energy constraints, developing effective security measures without overwhelming their limited resources is a significant challenge. Traditional security solutions, while robust, often impose heavy burdens on energy consumption, computational time, and overall complexity, making them impractical for energy-limited devices [42]. For instance, many cryptographic algorithms designed for high-security environments are not suitable for IoT devices due to their high power and processing requirements.

Recent advancements have explored analog-inspired encryption methods tailored for hardware security, which significantly reduce energy consumption, latency, and computational complexity. These methods, however, typically provide a moderate level of security, which might not be sufficient for all applications. Therefore, a critical aspect of securing energy-limited devices is finding a dynamic balance between security, power consumption, and performance. Adaptive security protocols that can scale their complexity based on the current operational context of the device are one promising approach. These protocols adjust the level of security in real time, ensuring that energy resources are utilized efficiently while maintaining an acceptable level of security [117]. This adaptive approach is particularly crucial in IoT ecosystems, where devices may operate under varying conditions and requirements. Establishing this dynamic trade-off is essential for the widespread adoption and trust in IoT devices, ensuring they can perform reliably and securely without compromising their operational longevity or functionality.

3.3 Confidentiality of Data/Data Privacy

Data privacy is a fundamental aspect of hardware security, critical for protecting sensitive information from unauthorized access and exploitation. The integrity of data confidentiality is paramount, especially when dealing with sensitive signals such as cryptographic keys, which, if compromised, can lead to severe security breaches. Traditional testing and validation methods often fail to identify vulnerabilities that arise from complex interactions within the hardware. This is particularly concerning as triggers for these vulnerabilities may not be embedded directly in the program code or data, making them elusive during standard testing procedures [125]. Consequently, ensuring data privacy necessitates comprehensive testing methodologies that can detect hidden triggers and other subtle threats.

Moreover, the increasing trend of outsourcing data storage to third-party cloud services introduces additional layers of risk to data security. While cloud solutions offer scalability and cost efficiency, they also pose significant challenges in ensuring the confidentiality of stored data. Data breaches in the cloud can occur due to various factors, including insider threats, insufficient access controls, and vulnerabilities within the cloud infrastructure itself. The reliance on third-party providers means that organizations must trust these providers to implement and maintain robust security measures, which is not always guaranteed. This scenario underscores the necessity for advanced encryption techniques, secure key management practices, and stringent access controls to safeguard data stored in the cloud. Additionally, adopting zero-trust architectures, where no entity inside or outside the network is trusted by default, can further enhance data privacy and security.

3.4 Side-Channel Leakage Reduction

Reducing side-channel leakage is a critical challenge in the domain of hardware security. Side-channel attacks exploit physical emissions such as heat, electromagnetic (EM) waves, noise, and vibration that occur as a by-product of hardware operations. These emissions can reveal sensitive information like cryptographic keys, which can be used to compromise the security of the entire system. As encryption is a common method for maintaining data confidentiality, its physical implementation can become vulnerability when subjected to side-channel attacks.

The threat posed by side-channel attacks arises from their ability to circumvent traditional security measures by targeting the physical layer of the hardware. For instance, an attacker can measure the power consumption or EM radiation emitted by a device during cryptographic operations to infer the secret keys used. These attacks do not rely on breaking the cryptographic algorithm itself but rather on exploiting the physical weaknesses in the implementation. As such, they pose a significant risk to the integrity of encrypted data.

Traditional hardware defenses against side-channel attacks, such as masking and hiding techniques, have been developed to mitigate these risks. Masking involves randomizing the intermediate values processed by the cryptographic algorithm to obscure the correlation between the physical emissions and the secret data. On the other hand, hiding techniques aim to reduce the observable side-channel signals by making the power consumption or EM emissions uniform regardless of the processed data. However, these methods often come with significant energy overhead, making them less feasible for energy-constrained devices like those in the Internet of Things (IoT) ecosystem. For example, implementing masking can substantially increase the power consumption and complexity of the hardware, while hiding techniques may require additional circuitry, further consuming valuable energy resources.

Innovative approaches are being explored to address these limitations and reduce side-channel leakage without incurring high energy costs. One such approach is the use of logic locking techniques, which obscure the hardware functionality by embedding additional logic gates that must be unlocked using a secret key. This method can effectively deter side-channel attacks by making the physical emissions less predictable and more challenging to analyze. Additionally, emerging technologies like cryogenic computing, which operates at extremely low temperatures, can potentially minimize thermal emissions and reduce the effectiveness of temperature-based side-channel attacks. Furthermore, research is also focusing on developing more energy-efficient masking techniques that can be applied dynamically, only when sensitive operations are being performed, thereby reducing the overall energy consumption. For example, adaptive masking schemes can selectively apply masking based on the sensitivity of the data being processed, balancing the trade-off between security and energy efficiency.

3.5 Flexible/Reconfigurable/Automatic Security

Balancing flexibility and performance is a critical challenge in the design of digital ICs, particularly in the context of security. ASICs offer minimal flexibility post-design but excel in performance for specific tasks. These highly specialized circuits are optimized for particular applications, delivering superior speed and efficiency with minimal power consumption. However, their fixed nature makes them less adaptable to new requirements or security threats that emerge after deployment. In contrast, reconfigurable computing, exemplified by Field-Programmable Gate Arrays (FPGAs), offers a high degree of flexibility. FPGAs can be reprogrammed to alter their architecture and functionality, allowing for dynamic adaptation to changing computational needs and emerging security threats. This reconfigurability makes FPGAs invaluable for applications requiring frequent updates or customization. However, this flexibility comes at the cost of reduced performance compared to ASICs, including slower operational speeds, higher power consumption, and larger physical area requirements.

From a hardware security perspective, the programmability of FPGAs introduces additional vulnerabilities. The increased attack surface results from the ability to reprogram the FPGA, potentially introducing malicious configurations or exploiting design flaws. For instance, unauthorized access to the reconfiguration interface could allow an attacker to alter the device's behavior, leading to data breaches or functional compromises. Furthermore, the reconfigurability could be exploited to bypass security mechanisms embedded within the hardware. Achieving a balance between flexibility and security in reconfigurable computing requires innovative design strategies. One approach involves developing secure configuration mechanisms that ensure only authorized updates are applied to the FPGA. This can be achieved through secure boot processes, cryptographic authentication of configuration files, and real-time monitoring for unauthorized changes. Additionally, partitioning the FPGA into secure and nonsecure zones can limit the impact of potential security breaches, ensuring that critical functions remain protected even if part of the device is compromised.

Automatic security mechanisms are another area of focus. These systems can dynamically adjust security settings based on real-time analysis of threats and operational conditions. For example, adaptive security protocols can strengthen encryption or enable additional protective measures in response to detected anomalies or attacks. Machine learning algorithms can be employed to continuously monitor the FPGA's behavior, identifying and mitigating potential threats before they can cause harm. Furthermore, the use of hardware redundancy and error-correcting codes can enhance the robustness of reconfigurable devices. By duplicating critical components and performing cross-checks, the system can detect and correct errors that may result from security breaches or hardware failures. This approach not only improves security but also increases the reliability and fault tolerance of the device.

3.6 Packaging-Related Challenges

In order to gain high performance, shorter time to market, cost efficiency, and small-sized ICs, advanced chip packaging is gaining attention, and as a result, packaging-related risks are also evolving nowadays. For example, fan-in and fan-out wafer-level packaging (WLP) offers more opportunities to manufacture high-performance and small-sized ICs [45]. System-on-Chip (SoC) and System-in-Packaging (SiP) technologies are cost-efficient approaches for the industry to follow Moore's law [19]. Famous chip foundries such as TSMC and Samsung are also involved in advanced IC packaging. Therefore, fabless production of advanced packaging is inevitable and is a significant source of vulnerability to the packaging process. Moreover, various alterations can be easily implemented due to globalization and the complexity of the microelectronics supply chain. On the one hand, the globalization of the IC supply chain makes it more challenging for end users or IP owners to guarantee that their components are guaranteed and tamper-free. On the other hand, the current detection methods cannot verify the internal materials of sample packaging due to the increased complexity of advanced IC packaging. In addition, adversarial changes in packaging parameters by a malicious manufacturer can result in undetected features, which can cause chip failure. As a result, chip packaging is now viewed as a severe hardware security issue [161].

3.7 Si-Backside Vulnerability

Data security has become a major issue because of the higher reliance on IoT edge devices that acquire confidential and private information and export it to the cloud network. Cryptography is used to provide network security for the IoT and edge devices by using two approaches: digital encryption algorithms for data protection and digital signature algorithms for device authentication. Several methods are used to safeguard this SoC module, for example, the front side of the SoC has metal layers and shielding meshes that block incident photons' paths and shield the ICs from optical attack. However, the silicon substrate of the gadget, which is its backside, lacks any such safeguards [155]. An attacker can effortlessly track and extract security-sensitive data and information by laser/optical probing, fault injection, and side-channel attacks. But, in practice, flip-chip packaging is frequently employed in these IoT devices because they provide a smaller form factor even with more functionality and lower power. A flip-chip exposes the Si substrate to the surface of an IC chip, leading to a considerable security gap. Flip-chip can be achieved with packaging that is narrow and practically chip-scale which makes them ideal for IoT devices. Therefore, an attacker can effortlessly track and extract security-sensitive data and information by laser/optical probing, fault injection, and side-channel attacks. As a result, a Si-backside of an IC chip, especially in flip-chip packaging, leads to a serious security hole [114].

Chapter 4
Security in Resource-Constrained IoT Devices

As the IoT continues to proliferate, the security of resource-constrained devices becomes a paramount concern. These devices, often operating with limited power, processing capabilities, and memory, are integral to a wide array of applications, ranging from supply chain management to healthcare. This chapter explores the security challenges specific to Radio Frequency Identification (RFID) systems, a key component of many IoT networks. RFID tags, due to their constrained nature, present unique vulnerabilities that demand lightweight and efficient security solutions. By examining the architecture, construction, and operational characteristics of RFID systems, as well as the various threats they face, this chapter highlights the critical need for robust security measures tailored to the limitations of these ubiquitous devices.

4.1 Security Challenges in Constrained RFID Systems

In the three-layer IoT architecture, the application layer forms the topmost layer of the IoT architecture which is responsible for effective utilization of the data collected. The physical layer, or the perception layer, comprises edge devices (IoT devices). Among these devices are RFID tags. RFID technology is used in various applications such as supply chain management, access control, identity authentication, and object tracking. The RFID tags are attached to "things" or objects to be tracked by RFID readers, as shown in Fig. 4.1. This is how things can gain connectivity and be integrated into smart IoT applications. RFID tags are the most constrained of IoT devices; consequently, the task of securing RFID tags is a most challenging endeavor. In this chapter, we take a close look at the RFID system and the security threats that are to be mitigated using lightweight approaches.

© The Author(s), under exclusive license to Springer Nature Switzerland AG 2025
K. Khalil et al., *Lightweight Hardware Security and Physically Unclonable Functions*, https://doi.org/10.1007/978-3-031-76328-1_4

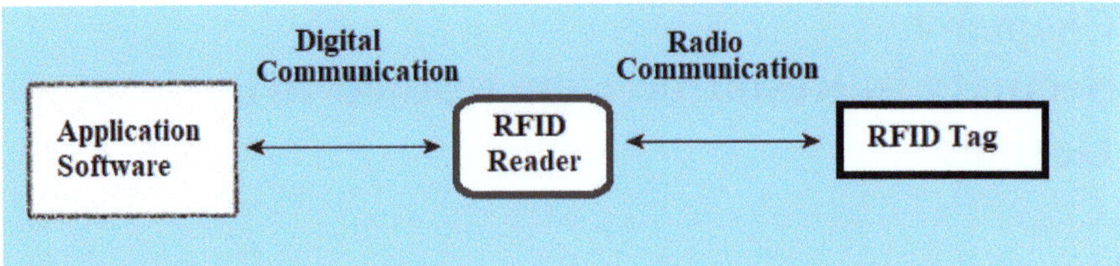

Fig. 4.1 A typical RFID system

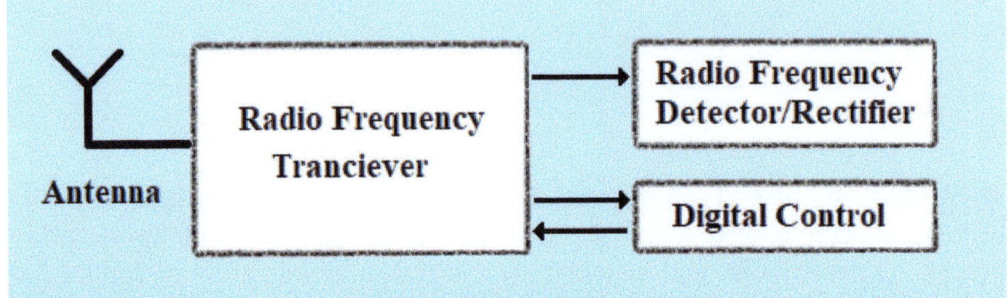

Fig. 4.2 Block diagram of a passive RFID transponder

4.2   RFID System Architecture

RFID is a form of automatic identification and data capture (AIDC) technology [76].
An RFID system is used to identify many types of objects, such as manufactured
goods, livestock, and people. An object that needs to be identified has an RFID tag
attached to or embedded within the object. A typical RFID system comprises an
RFID transponder (or an RFID tag), an RFID interrogator (or an RFID reader),
and an application software that acts as an interface between the user and the
RFID system [124]. Each tag has a unique identifier and might also hold additional
information about the object. RFID readers communicate wirelessly with the tags
to identify and track the corresponding objects and devices connected to each tag.
Depending on the application and tag's capabilities, RFID readers might also be
able to read and update additional information stored on the tag.

The main purpose of an RFID tag is to communicate its ID and useful
information about the object it is attached to. The block diagram of a passive RFID
transponder is shown in Fig. 4.2, and it is composed of:

- An antenna
- A radio frequency (RF) transceiver section
- An analog detection and/or rectification section, which detects and, in passive
  tags, rectifies RF power into an equivalent dc voltage
- A digital control section that is either a microprocessor or some other digital
  system

Fig. 4.3 Construction formats of RFID tags

4.3   RFID Construction Formats

RFID technology is incorporated in many applications, each requiring different physical construction properties of the transponders. Common tag construction formats include disks or coins, glass or plastic housing, keys and key fobs, smart labels, coil-on-chips, and those that are embedded in smart cards [46]. The various construction formats are summarized in Fig. 4.3 [58].

4.4   RFID Tag Characteristics

The market for RFID tags includes many different types of tags, but they differ greatly in size, cost, functionality, and security. Tags are usually designed with certain characteristics to comply with a particular standard and are often further customized to meet the requirements of specific applications. Understanding the major tag characteristics can help those responsible for securing RFID systems

Table 4.1 Comparison of passive, semi-passive, and active tags

Tag type	Passive	Semi-passive	Active
Power source	Harvesting RF energy	Battery	Battery
Communication	Respond only	Respond only	Respond or initiate
Max range	10 M	>100 M	>100 M
Relative cost	Least expensive	More expensive	Most expensive
Example	EPC	Electronic tolls	Large-asset tracking

identify the best security solutions to implement on RFID tags to satisfy their needs and ensure efficient use of the tag's resources.

4.4.1 Power Supply

RFID tags require power to perform different functions, such as sending data to a reader, retrieving and storing data, and performing other computations demanded by the application and security needs. Tags can acquire this power either from a battery or from electromagnetic waves emitted by the RFID readers. A tag's power requirements depend on several factors, including the expected operating distance between the tag and the reader, the operating radio frequency in use, and other functionalities that might be required of the tag. The more complex the functions an RFID tag supports, the higher its power requirements. Therefore, tags that support encryption and/or authentication require more energy than tags that are only limited to transmitting their ID. RFID transponders are classified based on their power supply requirement [124, 158]. There are three types of tags, which are summarized in Table 4.1 [158] and can be defined as passive, active, or semi-passive tags.

4.4.1.1 Passive Tags

Passive tags do not possess an onboard power supply and thus are unable to initiate communication. Passive tags rely only on the power emitted from the reader for both data processing and transmission. Due to the lack of an onboard power supply, passive tags have a much shorter reading range than active or semi-passive tags. Passive tags are more sensitive to environmental noise and interference; they have little or no data processing capabilities, and for this reason, they cannot be easily reprogrammed. While passive tags have such shortcomings, they are the cheapest to manufacture and the easiest to integrate into products because of their small size. Such features allowed passive tags to be used in a wide range of applications, such as medical applications, supply chain management, and wireless sensing.

4.4.1.2 Active Tags

Active tags have an onboard power source, such as a battery, and may initiate communication with a reader or other active tags. Active tags often have a much longer operating range than passive and semi-passive tags and can process and store more data. Active tags are often used to track large assets and livestock, as the items they are attached to are often high in value and allow for bulkier tags.

4.4.1.3 Semi-passive Tags

The main difference between an active tag and a semi-passive tag is that a semi-passive tag has an onboard power supply dedicated to only signal processing tasks and is not utilized to amplify the received and transmitted signals. Semi-active tags cannot initiate communications; they must first acknowledge the interrogating signal of the reader to reply. This communication protocol is known as interrogator-driven. A semi-passive tag consumes much less power from the onboard battery and has a longer life than an active tag. Semi-passive tags can also perform complex data processing and encryption tasks and achieve the reading range almost as good as active tags.

4.4.2 Functionality

A tag's primary function is to provide an identifier to a reader, but many types of tags can support additional capabilities that are required for different applications. An RFID tag can differ in terms of on-chip memory, embedded sensors, and security features [76]. Nonvolatile memory enables data to be stored on tags and retrieved later; this enables more flexibility in the design of RFID systems as it allows for data transactions without concurrent access to the data stored on a separate server. Depending on the application, additional volatile memory might be required to execute various computations usually associated with the provided security features. Environmental sensors with tags are an example of the benefit of adding local memory, which allows a tag to record sensor readings to be later retrieved by a reader. Tags with onboard memory are often coupled with security modules to protect the stored data and provide authentication between the tag and the reader. Security features can range from supporting a lock command that can prevent further alteration or access of the data on memory to supporting advanced cryptographic algorithms that enable authentication and tamper protection. However, the integration of additional memory, sensors, and security on tags significantly increases the tags' cost, power requirements, and complexity.

4.5 RFID Security Threats

As the RFID uses wireless means of communication between the reader and the tag, the RFID systems may be faced with eavesdropped, counterfeiting, playback, and tracking threats. This brings up communication security concerns, such as privacy leaks and unauthorized [95, 116]. As there are so many security threats for RFID, attacks can be classified into the physical, system, and channel threats according to the objects and means.

4.5.1 Physical Threats

A physical threat is a threat that uses physical means to attack labels or communication systems, such as reverse engineering and physical exploration. To maintain a low cost, most RFID tags will not be equipped with a tamper-resistant mechanism. An attacker with physical access to a tag can duplicate a tag through reverse engineering, and physical probing is capable of getting confidential information stored within the tag.

4.5.1.1 Tag Cloning

RFID used for identification has a unique ID number; if the ID information is exposed, an adversary can easily clone the tag. Now that many programmable read-write tags are put into use, cloning a tag would not be difficult.

4.5.1.2 Passive Interference

Because RFID networks operate in an inherently unstable and noisy environment, their communication is rendered susceptible to possible interference and collisions from any source of radio interference, such as noisy electronic generators and power switching supplies. This interference would prevent accurate and efficient communication.

4.5.1.3 Active Jamming

Passive interference is usually unintentional; however, an adversary is capable of causing electromagnetic jamming by broadcasting a signal within the same range as the reader to prevent tags from communicating with other readers.

4.5.1.4 Tapping

Because RFID readers and tags communicate wirelessly, they are vulnerable to eavesdropping. Eavesdropped information is often used to achieve other more elaborate attacks, such as counterfeiting and tracking. Since the transmission power of the reader is much larger than that of a tag, eavesdropping on the reader is more accessible and can be done at greater distances. The communicated data between a reader and a tag must be encrypted to prevent confidential information from being eavesdropped.

4.5.1.5 Replay Attack

One of the most severe threats which RFID systems face is the replay attack. The replay attack is when a malicious node or a device replays critical information eavesdropped through communication between reader and tag to achieve unauthorized access. A typical application is when the illegal device plays back the authentication between the reader and the tags, deceiving readers or tags to pass verification. Solutions to replay attacks include using a one-time password authentication protocol or updating ID information dynamically.

4.5.1.6 Relay Attack

A relay attack, or a man-in-the-middle attack, is when an attacker places a malicious device between the reader and the tag to intercept the information between two nodes and then alter its content or forward it directly to another device.

4.5.2 System Threats and Side-Channel Attacks

System threats mainly refer to the attacks targeting the flaws in the authentication protocol and encryption algorithm. In contrast, side-channel attacks are when attackers bypass the communication of an RFID system through the use of "Timing and Power Analysis" attacks. Solutions for preventing side-channel attacks include reducing the electromagnetic radiation, enhancing the complexity of the algorithm of the RFID chip, making it difficult for the attacker to understand the internal encryption algorithm information.

4.5.2.1 Counterfeiting and Spoofing Attacks

Attackers get some information about identity by detecting communications between readers and legitimate tags. Counterfeiting or spoofing is when eavesdropped identity information is used to impersonate genuine labels or readers. An attacker can fake labels, as well as fake readers. An effective strategy to prevent counterfeiting and spoofing attacks is to use efficient two-way authentication protocols between tags and readers.

4.5.2.2 Tracking

The tags used as identification have unique identity information (such as ID number), so they are easily tracked. Tracking allows adversaries to extract secret and private information about the user, such as activity patterns and hobbies. To defend against tracking, the ID information sent by the tag each time should be dynamic. A common approach used is to utilize a pseudorandom function to update a tag's ID dynamically.

4.6 PUF-Based RFID Security Solutions

The unclonable, lightweight, and tamper-resistant nature of PUF primitives makes them a suitable choice for improving the security of RFID tags. PUF can be used as digital fingerprints that help securely identify tags or as secure key generation used in PUF-based authentication protocols [151]. However, due to the shortcomings of many PUF primitives, such as low reliability, modeling attacks, and small challenge-response space, achieving reliable security remains challenging. Mitigating the shortcomings of PUF primitives requires additional area overhead, computation complexity, and power consumption, which prohibits their practical use in RFID systems. For this reason, careful consideration of the choice of PUF primitives and the need for tailored PUF security designs and protocols are required to satisfy various applications.

Chapter 5
Physically Unclonable Functions

Significant research effort has been put into producing lightweight security protocols that can be implemented in constrained devices. Typical lightweight security utilizes:

- RNGs
- Simple functions like CRC checksums
- Simple bit operations

However, protocols that only utilize lightweight elements were found to have varying degrees of vulnerability against dedicated attackers. As such, constrained devices are still unable to employ reliable security protocols due to the prohibitive power and implementation area costs of cryptographic security primitives. PUFs [4, 54] or Physically Unclonable Functions are emerging hardware security primitives that offer a lightweight alternative to standard security. In this chapter, we examine the PUF as new hardware security primitives and explore their use in lightweight authentication protocols.

5.1 Overview of Physical Unclonable Functions

A PUF is a physically disordered system that reacts to an external challenge C with a unique response R [63] as shown in Fig. 5.1. Although the response cannot be predicted from the circuit, it is reliable, i.e., a challenge C will always produce the same response R when applied to the same circuit. Applying the same challenge C to a different circuit would produce a different repose even if both PUF circuits share the same design and layout. Hence, the PUF design and layout are not secret, and the unclonability relies on the minuscule variations of the physical properties of the device during fabrication. In a sense, a PUF functions as a cipher and can

K. Khalil et al., *Lightweight Hardware Security and Physically Unclonable Functions*, https://doi.org/10.1007/978-3-031-76328-1_5

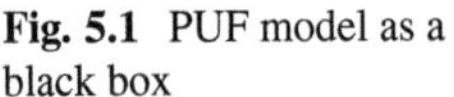

Fig. 5.1 PUF model as a black box

be used to transform a plain text C to a cipher text R. However, there is one key difference in the PUF when compared to standard encryption methods: There is no key.

With the absence of the need for a key, a great security weakness is removed. Key storage would expose the system to a new vulnerability: key extraction. Once extracted, all communication can be intercepted. This eliminates the need for secure storage and active intrusion detection as PUFs are naturally resistant to probing due to their high sensitivity.

5.1.1 PUF Quality Metrics

While a PUF is physically unclonable, it can still be modeled mathematically if it has low entropy. There are two important quality metrics that related to the PUF [87]:

1. Inter-chip Hamming Distance quantifies the difference between manufactured PUFs of the same design and layout and thus represents the degree of randomness achieved through fabrication. For ideal randomness, each bit of the PUF response output would have a 50% chance of having the same value of that produced by another PUF utilizing an identical layout.
2. Intra-chip Hamming Distance quantifies the variability of response of a single PUF when represented with the same challenge multiple times. Ideally, it should be null. In practice, it is often not the case, and there is a portion of the PUF output that is unstable. Error-correcting circuits are used to ensure correct responses. A PUF needs to reliably produce the same response R when it is represented with a challenge C. PUFs with a nonnegligible intra-chip hamming distances are often used in conjunction with error detection and correction circuits.

5.1.2 Strong PUFs Versus Weak PUFs

The challenge-response space of the PUF determines whether it is classified as a weak PUF or a strong PUF [63]. Strong PUFs have a large challenge-response space, which allows them to be used in challenge-response authentication protocols. Strong PUFs are able to provide a complete authentication solution for a device. Such authentication is very similar to how secret-question/secret-answer authentication is performed. This is illustrated in Fig. 5.2.

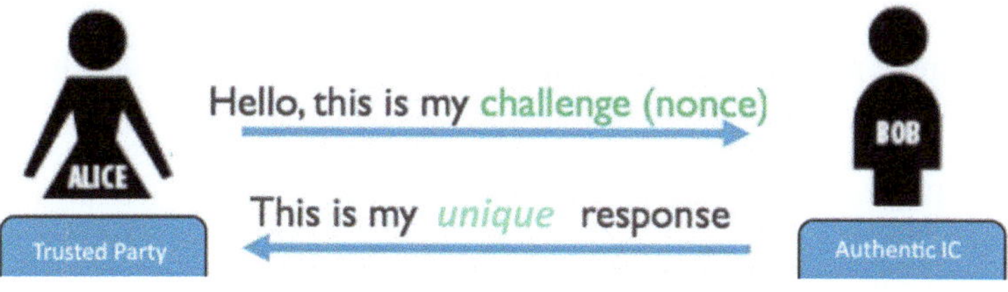

Fig. 5.2 Alice, a trusted party, authenticating a PUF device, Bob, using challenge-response authentication

Alice would send Bob a challenge. Bob would apply the challenge as an input to the PUF circuit to produce a response. Bob would send the produced response to Alice where she matches it with a database of previously collected challenge-response pairs. If it is a match, Bob's identity is deemed authentic. Since the challenge-response pair is exposed, it should never be used again. By using each challenge once, we do not run the risk of having a recorded response used by an untrusted party.

Weak PUFs are characterized by having a small challenge-response space. Such PUFs are used as an alternative to secure key storage. By having a hidden response, these PUFs will reliably generate an agreed-upon key. This key will never have the need to be stored offline since we are able to reliably generate it from the PUF circuit during run time. By using weak PUFs, we save the circuitry and power needed to store and load the secret key. Furthermore, the chance of the key getting extracted is extremely small, as any tampering in the device will destroy the PUF model used for key generation. The PUF circuit relies on tiny fabrication variations to produce its output; therefore, any tampering attempt is highly likely to alter these variations, making it difficult for a third party to extract or generate the secret key.

5.2 Challenge-Response Pairs Management

To allow for authentication, the trusted party must be able to confirm whether a PUF response is authentic or not. Since the trusted party would not have access to the PUF circuit, it should have other means for storing or producing an authentic response. After the fabrication of a PUF, challenge-response pairs should be recorded and stored at an authentication server. Since each challenge should be only used once, enough pairs should be stored for the expected lifetime of the PUF. This requires a very high storage capacity when dealing with systems of thousands of devices, which is very reasonable.

A more reasonable approach is to build a software model of the PUF circuit by performing a machine learning modeling attack. The soft model would be able to generate a response for any challenge, and there would be no need to store a large number of challenge-response pairs. However, a model-based approach might be problematic for PUFs that are resilient to modeling attacks. To enable such attack, the PUF circuit must be probed prior to packaging to extract additional information that would allow for modeling.

Although this eliminates the need of storing challenge and response pairs, the model itself needs to be stored securely to secure against attacks that might leak the model into the hands of untrusted parties. Another caveat is the fact that the PUF needs to be vulnerable to probing attacks for this method to work. This introduces the need for protection against invasive attacks to prevent untrusted parties from probing the circuit in order to arrive at the model.

Chapter 6
PUF: Security Threats

PUFs serve as robust security primitives due to their inherent resistance to traditional cryptographic attacks. However, understanding and mitigating potential security threats against PUFs is crucial for ensuring their effectiveness in practical applications.

6.1 Machine Learning Attacks

A strong PUF's resistance to modeling attacks is the cornerstone behind PUF security. This resistance is measured in NCRPs, the number of authentic challenge-response pairs that need to be observed by an attacker to achieve certain prediction accuracy. Most successful attacks on Strong PUFs usually utilized machine learning algorithms such as Deep Neural Networks (DNNs), Support Vector Machines (SVMs), and Evolutionary Strategies (ESs). Ideally, an attacker would not be able to predict the response of any future challenges inputted into the circuits regardless of the NCRPs observed.

6.2 Side-Channel Attacks

A side-channel attack can be made by statistically analyzing the circuit's power consumption, time delays, or electromagnetic (EM) emission to gain knowledge about its secrets. Merli et al. [112] performed a side-channel analysis on RO-PUFs, showcasing that the frequencies of ROs on FPGA can be measured using EM equipment. Attacks based on a combination of side-channel analysis and machine learning attacks have shown to be successful on arbiter PUF design variants, which are more resistant to conventional machine learning attacks [35]. The assumption

© The Author(s), under exclusive license to Springer Nature Switzerland AG 2025
K. Khalil et al., *Lightweight Hardware Security and Physically Unclonable Functions*, https://doi.org/10.1007/978-3-031-76328-1_6

that PUFs are generally more resistant against side-channel attacks is not always true, and careful consideration of side-channel resistance needs to be considered when evaluating the security PUF primitives.

6.3 Environmental Attacks

PUFs are susceptible to environmental factors such as temperature variations, electromagnetic interference, and aging effects, which can degrade their reliability and security. Temperature fluctuations can alter the characteristics of PUF components, leading to variations in response behavior. Electromagnetic interference from nearby electronic devices can introduce noise into PUF signals, compromising their integrity. Additionally, prolonged use of PUFs can result in aging effects, causing drifts in response patterns over time. Mitigating environmental attacks requires robust design techniques, such as temperature compensation mechanisms and shielding against electromagnetic interference, to ensure consistent and reliable PUF operation.

6.4 Fault Injection Attacks

Fault injection attacks involve deliberately inducing faults in a PUF circuit to manipulate its behavior and extract sensitive information. Common techniques for fault injection include voltage and clock glitching, laser-induced fault injection (LIFI), and electromagnetic fault injection (EMFI). By exploiting vulnerabilities in the PUF circuitry, attackers can perturb its operation and extract responses that reveal its secrets. Countermeasures against fault injection attacks include redundancy-based error detection, error correction codes, and physical tamper resistance techniques to thwart unauthorized access to PUFs.

6.5 Hardware Trojans

Hardware Trojans are malicious modifications inserted into PUF circuits during fabrication or post-fabrication stages with the intent of compromising their security. These Trojans can manifest as additional logic gates, hidden functionalities, or modifications to existing circuitry. Once activated, hardware Trojans can leak sensitive information, degrade performance, or even render PUFs unusable. Detecting hardware Trojans in PUFs requires thorough testing and verification methodologies, such as hardware security audits, side-channel analysis, and circuit-level inspection, to ensure the integrity and trustworthiness of PUF implementations.

Chapter 7
PUF Performance Testing and Evaluation

PUFs offer a promising approach to securing devices by leveraging the inherent physical variations in hardware. However, to ensure the efficacy and practical application of PUFs, comprehensive performance testing and evaluation are crucial. This chapter delves into the key performance metrics that are essential for quantifying the effectiveness and reliability of PUF circuits. These metrics, including randomness, uniqueness, and reliability, are critical in assessing how well a PUF can resist modeling attacks, maintain consistency under varying environmental conditions, and provide distinct responses across different instances. Additionally, the chapter examines the importance of area, power, and delay metrics, which determine the feasibility of integrating PUFs into resource-constrained devices like RFID tags and sensor nodes. By rigorously evaluating these parameters, we can better understand the capabilities and limitations of PUFs in real-world applications, guiding the development of more robust and efficient security solutions.

7.1 PUF Performance Metrics

While a PUF is physically unclonable, it has been shown that it can be mathematically modeled, making it prone to modeling attacks. Furthermore, PUF responses are usually unreliable due to their sensitivity to thermal noise, varying environmental conditions, and aging. The following are important metrics used to quantify and evaluate the performance of a PUF circuit:

© The Author(s), under exclusive license to Springer Nature Switzerland AG 2025

K. Khalil et al., *Lightweight Hardware Security and Physically Unclonable Functions*, https://doi.org/10.1007/978-3-031-76328-1_7

7.1.1 Randomness

Randomness indicates the random guessing predictability of a PUF response. A PUF's responses should be unpredictable; for this to be true, the probability of having a response bit of "0" or "1" to a random challenge should be identical. Let P1 be the bias of responses toward the value "1". Randomness (H) of a PUF device is then evaluated. An Ideal PUF would have a Randomness H = 50, where H is defined as follows:

$$H = \min(P_1, 100 - P_1), \quad \text{where } 0 \leq H \leq 50. \tag{7.1}$$

7.1.2 Uniqueness

Uniqueness is a measure of identifiability, where responses from different PUF instances should be distinguishable. The uniqueness of a PUF is measured by calculating the Inter-chip Hamming Distance (Inter-chip HD). Let D be the normalized Inter-chip HD between two PUF instances. An ideal PUF would have a Uniqueness U = 50, where H is defined as follows:

$$U = \min(D, 100 - D), \quad \text{where } 0 \leq U \leq 50. \tag{7.2}$$

7.1.3 Reliability

A reliable device is expected to produce the same response if repeatedly given the same challenge. We assess the reliability of a device based on its bit error rate (BER), which represents the percentages of response errors observed under varying operating conditions.

7.1.4 Area, Power, and Delay Metrics

The design of PUF primitives focuses on critical performance metrics such as area, power, and delay, which determine their applicability in various devices and applications. Although these metrics are often considered secondary to the PUF's security, many of the PUF applications impose strict area, power, and/or throughput requirements. Depending on the application and the device's available computation resources, the choice of PUF and its design is tailored to satisfy such constraints. The area overhead of a POUF design directly impacts the cost of the die, and a small area enables the use of PUF primitives in low-cost and constrained devices such as RFID tags and sensor nodes. Higher throughput of response generation may also be required for critical health applications, where delays are not tolerated. This work evaluates such PUF performance metrics for use in lightweight devices.

7.1.5 Sensitivity to Environmental Conditions

PUF responses can be affected by environmental factors such as temperature, humidity, and electromagnetic interference. Evaluating the sensitivity of PUFs to these conditions ensures their robustness and reliability in real-world deployment scenarios. PUFs that exhibit minimal variations in response under different environmental conditions are preferred for applications requiring consistent performance.

7.1.6 Resistance to Aging

Over time, PUF responses may degrade due to aging effects caused by factors such as device wear-out and material degradation. Assessing the long-term stability of PUFs and their resistance to aging ensures the durability and reliability of security solutions deployed in the field. PUFs that maintain consistent performance over extended periods are preferable for applications requiring longevity and reliability.

Chapter 8
Improving the PUF's Reliability

The reliability of PUFs is a critical aspect of their performance, particularly in environments with varying thermal, electrical, and aging conditions. This chapter focuses on strategies to enhance the reliability of PUFs, ensuring consistent and error-free responses under diverse operating conditions. By addressing the inherent sensitivity of PUFs to environmental fluctuations and noise, we can significantly improve their practicality and robustness in security applications. Techniques such as error correction codes, environmental compensation methods, and advanced PUF architectures are explored in detail. These approaches aim to mitigate the bit error rate (BER) and ensure that PUFs can reliably generate the same responses when presented with the same challenges, thereby strengthening their role in secure authentication and key generation systems.

8.1 Delay-Based APUF: Security and Reliability

The APUF, with its small area overhead and large challenge-response space, makes for an attractive choice for secure low-cost, constrained devices. However, it suffers from two major shortcomings: the low reliability of its responses and the fact that it can be easily modeled by an adversary if the challenge-response exchange is observed. Those challenges prevent the APUF from being used for authentication without added modifications and proper security measures.

8.2 APUF's Poor Reliability

The main problem the APUF suffers from is the inherent low stability of its generated responses. Due to how an APUF operates, often the delay difference

© The Author(s), under exclusive license to Springer Nature Switzerland AG 2025
K. Khalil et al., *Lightweight Hardware Security and Physically Unclonable Functions*, https://doi.org/10.1007/978-3-031-76328-1_8

(Δt) between the two paths is very small and sensitive to noise and operating conditions. Varying operating conditions, such as temperature and supply voltage changes, aging effects, and random noise, would induce a change in the measured delay difference, Δt, and potentially alter the PUF's response. Consequently, faulty responses generated by APUFs, whether it is because of aging or temperature and voltage variations, are directly related to the measured delay difference Δt of the two compared paths [162, 171]. The smaller the delay difference Δt associated with a certain challenge, the more it is prone to error. We confirmed this finding by building a soft model of an APUF based on the delay variations of 65 nm CMOS technology [139]. We then introduced random noise to its delay component, and delay difference Δt of 10,000 challenges was observed along with the PUF's response, as shown in Fig. 8.1, where Δt of challenges associated with faulty responses that are centered at $|\Delta t| = 0$ and fall between $\Delta t = -36.1$ ps and $\Delta t = 36.7$ ps. Due to its poor repeatability, the traditional APUF is not suited for use in security applications such as secret key generation. Additionally, the APUF's security against modeling attacks is further diminished due to the efficiency of reliability-based CMA-ES attacks [35] in successfully modeling the APUF by exploiting sensitive delay information revealed by its unstable responses.

An APUF-based device would drastically increase reliability if we could guarantee that its responses have a large $|\Delta t|$. However, it is infeasible to directly measure the Δt of the two paths as the use of fine-resolution Time-to-Digital Converters

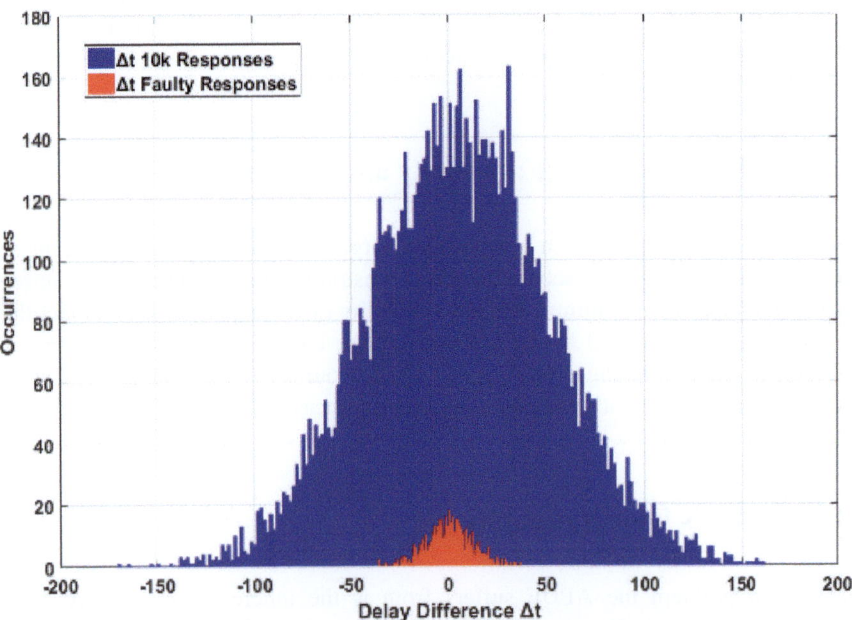

Fig. 8.1 Delay difference Δt (*ps*) of 10k responses (blue), and the subset of flipped responses (red) at BER $= 9.8\%$

(TDC) is unsuitable for constrained devices due to their large computational complexity and area overhead [94]. In this chapter, we introduce a new delay-based PUF with a novel architecture that is capable of guaranteeing a large delay difference $|\Delta t|$ at its output, thus drastically increasing the reliability of the PUF's responses while also utilizing a very small circuit implementation area that is suitable for constrained devices. Numerous enhancements have been made to the basic Arbiter PUF construction focused on enhancing the reliability of its responses. Most often, providing high reliability is accompanied by increased area overhead and computational complexity.

8.3 Reliability Enhancing Approaches

The delay-based APUF, characterized by its simple silicon-based design, small area footprint, and its exponentially large challenge-response space, has seen growing research interest as a promising security solution for constrained, low-power devices [98, 133]. However, one of the major shortcomings of the APUF that hinders its wide adoption in security applications is its inherent low reliability. Due to the nature of APUFs, their delay-based responses are highly unstable when working under different operating conditions. In an effort to alleviate this problem, several research efforts have targeted enhancing the reliability of the APUF. In this section, we discuss some of the different approaches that have been used to enhance the reliability of the APUF's generated responses.

8.3.1 Error Correction Codes

Error-Correcting Codes (ECC), such as BCH codes [18] and Fuzzy Extractors [102], are common techniques used to recover errors introduced in PUF's responses. In most cases, using ECC comes at the cost of imposing large implementation area overhead and/or a high computational complexity, which scales up quickly as the number of errors expected to be recovered increases [14]. Another drawback of using ECCs is the leakage of information associated with the ECC's helper data, which can be exploited to model the PUF [36]. Nevertheless, research efforts enhancing the APUF reliability would also drastically improve the efficiency of the ECC implementations as ECC implementation would be much less computationally expensive when addressing lower error rates. Index-Based Syndrome (IBS) [165] is a unique error correction code that is characterized by its high coding gain and reduced error correction complexity. IBS is best suited for PUFs with real-valued or "soft-decision" responses, such as that of RO PUFs and k-sum PUFs. However, IBS has been shown to leak security information when used on k-sum PUFs, due to the high correlation between the k-sum PUF's CRPs [11].

8.3.2   Temporal Majority Voting

Temporal Majority Voting (TMV), or Repetition Coding, is one of the lightweight approaches that can also be used to improve the reliability of the PUF's responses [163]. When using TMV, multiple evaluations of the same challenge are performed on the PUF device, and the final response bit would be decided by running a majority vote on the evaluated responses. In [163], TMV combined with BCH coding is used to improve the reliability of an APUF. The TMV is first applied to lower the bit error rate of the APUF's response before applying BCH. This combination would require the BCH implementation to have a much smaller area overhead and a lower computational complexity. However, TMV suffers from two major problems when used to improve the reliability of the APUF's responses. Highly unstable responses with $|\Delta t| \approx 0$ are equally likely to have a value of "0" or "1," and TMV will fail in suppressing resulting errors. Another problem TMV suffers from is the inability to detect and rectify errors resulting from the PUF being operated under nonoptimal temperature and voltage conditions, as resultant delay variations would be non-transient.

8.3.3   Selective Response Filtering

Selective challenge filtering is another approach used to improve the reliability of an APUF's generated responses. The server would utilize a stored software model of the APUF and manually select reliable challenges by inspecting their output delay difference [52, 159, 162]. Reliable challenges are chosen based on the software model's estimation of the underlying delay difference of the challenge. A delay difference threshold $|\Delta t_{t_h}|$ is then used to identify whether a challenge is reliable or not. Challenges that would have a $|\Delta t| > |\Delta t_{t_h}|)|$ would be considered reliable and would be selected for use. The value of the $|\Delta t_{t_h}|$ varies based on the process variation technology and the value of the BER that needs to be achieved. The higher the value of $|\Delta t_{t_h}|$, the lower the BER is. This comes at the cost of discarding a large subset of the CRP space available. This approach is highly effective since the added computational complexity on the server side is negligible. However, the reliability improvement is restricted to challenges generated on the server side, and an APUF device would not be capable of generating its own reliable challenge-response pairs. This approach would allow the server to authenticate the PUF device and send the device robust security keys utilizing the reliable challenge selection. However, the PUF device would still be unable to generate reliable challenges to authenticate the server. As a result, mutual authentication would still not be achievable without the use of costly ECCs.

8.3.4 Subthreshold CMOS Technology

Another circuit level, reliability enhancing technique is the proposed 45nm subthreshold arbiter PUF Design [96], which achieves a low power, high uniqueness, and improved reliability levels across varying operating conditions (a reported worst-case BER of 1.9%). One drawback of the subthreshold arbiter PUF design is the low throughput performance, which should be taken into consideration when employing such a design in critical applications. Our proposed work introduces a new PUF design that can achieve a BER of at most 0.98% when operating in extreme conditions. The newly introduced PUF would have a very small implementation area overhead, while it can also benefit from all previously mentioned reliability enhancing techniques as it operates at a different architectural level.

8.4 Vulnerability to Modeling Attacks

Another major shortcoming of the delay-based APUF is its poor resilience and vulnerability against machine learning (ML) modeling attacks. In such attacks, adversaries would collect the exchanged challenge-response pairs (CRPs) used in authentication sessions and apply machine learning algorithms to produce a software model of the PUF circuit. Such a model would be capable of correctly predicting the response of any incoming PUF input challenge. The Arbiter PUF can be modeled as a set of delay elements. The delay difference at the arbiter Δ can be expressed as a function of the differential delay vector ω, and Θ the feature vector that is a function of the input challenge [131]:

$$\Delta = \omega^T \Theta, \tag{8.1}$$

$$where, \ \Theta = \prod_{i=1}^{k} (1 - 2b_i). \tag{8.2}$$

In order to increase the non-linearity of the APUF delay model, different approaches are purposed in the literature in order to prevent successful ML-based attacks. In this section, we review some of the most popular approaches proposed to the security of the APUF.

8.4.1 XOR PUF

One of the most common methods to add non-linearity to a PUF design is the XOR PUF. In a k-XOR PUF, k Arbiter PUFs are placed on the chip. Each of the Arbiter PUFs receives the same challenges, and the responses of the n PUFs are XOR-ed

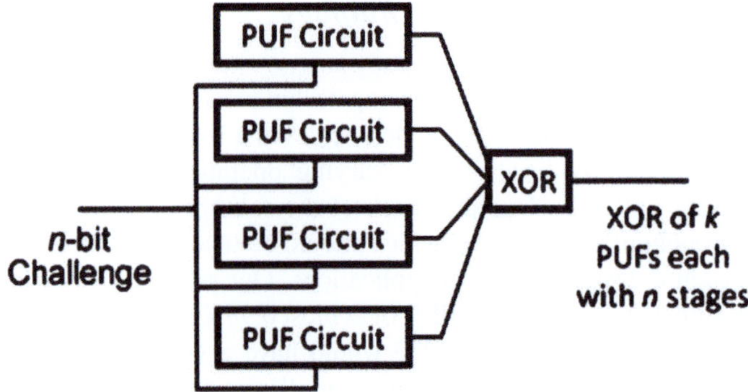

Fig. 8.2 Design Architecture of an XOR PUF

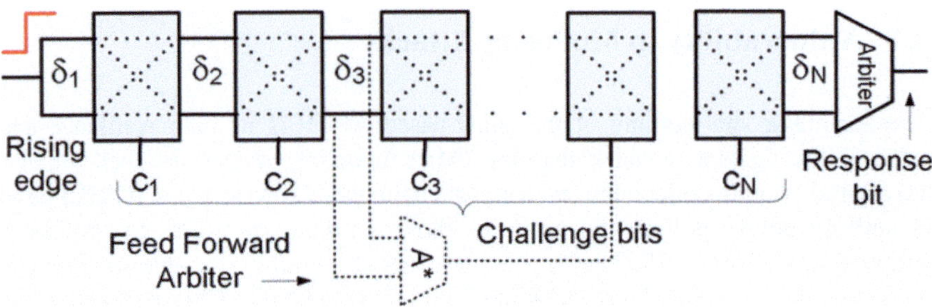

Fig. 8.3 Feed-forward PUF architecture

to build the final response bits, as shown in Fig. 8.2. While the machine learning resistance increases by XOR-ing more PUFs, adding additional PUF instances obviously also increases the area overhead of the design. Furthermore, the XOR PUFs become more unreliable, the more PUFs are XOR-ed. This limits the number of XORs that can be used in practice.

8.4.2 Feed-Forward PUF

The Arbiter PUF offers a strong PUF construction achieving a high degree of uniqueness and reliability. However, it is susceptible to software modeling attacks. In order to overcome this limitation, feed-forward Arbiter PUFs were proposed. In this PUF construction, some of the challenge bits are generated by intermediate signals along with the PUF structure as shown in Fig. 8.3. This introduces non-linearity in the design, making it infeasible to build a software model based on linear additive delay assumptions. Another approach to overcome software modeling

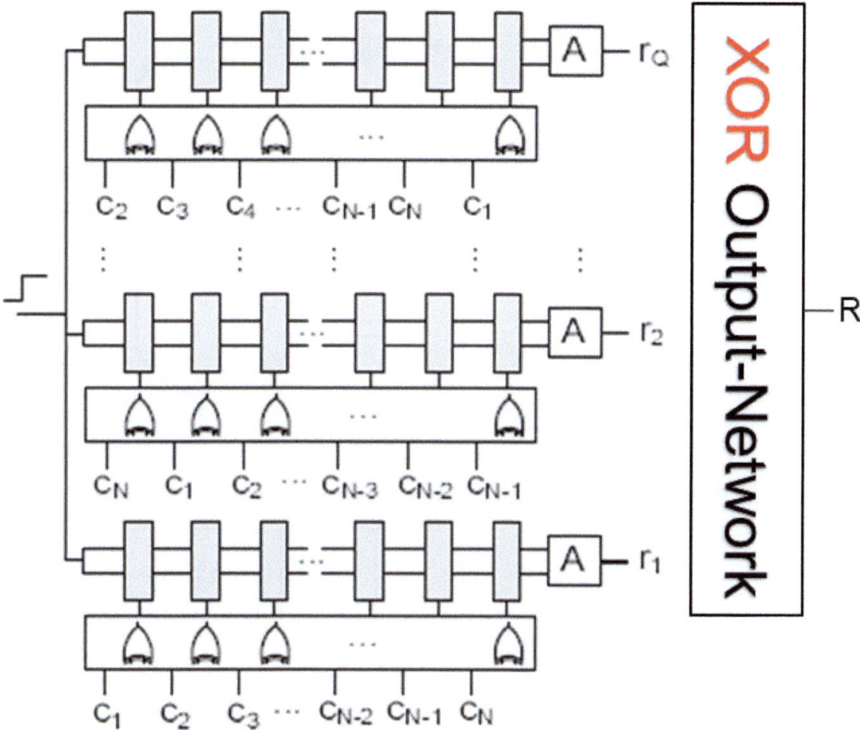

Fig. 8.4 Lightweight secure PUF architecture

attacks is to obfuscate the output response bits of the PUF. XOR Arbiter PUFs achieve this by XOR-ing the output response bits of multiple PUFs. An alternative approach would be to use the PUF response bits as challenge bits to a second PUF.

8.4.3 Lightweight Secure PUF

Lightweight Secure PUFs, shown in Fig. 8.4, incorporate four building blocks, namely, an input logic network, output logic network, an interconnect network, and parallel Arbiter PUFs to come up with a robust PUF circuit. The three main features of this PUF are:

- Inclusion of multiple delay lines to generate response bits
- A combination of input challenge bits using the input logic network
- A combination of PUF outputs through output logic network

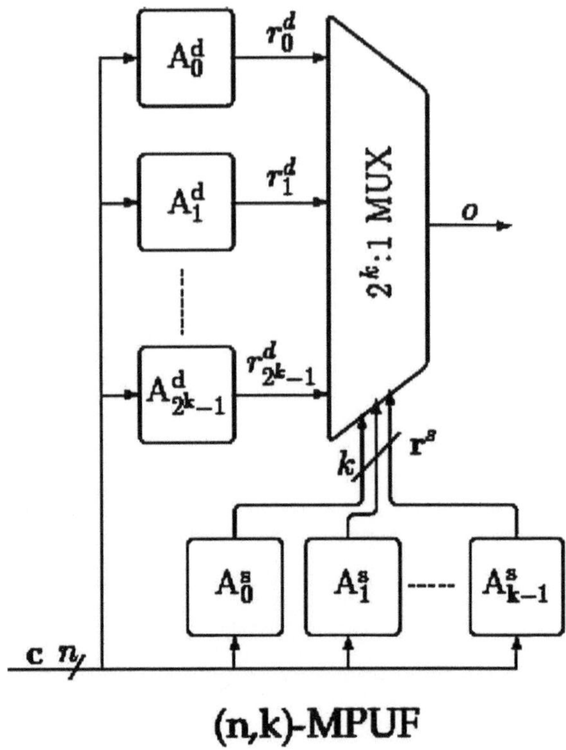

Fig. 8.5 A (n,k) multiplexer-based arbiter PUF architecture

The input network is constructed through XOR gates to create different combinations of challenge bits for each of the PUFs. The output logic network also consists of XORs to combine responses from different PUFs. The lightweight, secure PUF circuit is resistant to reverse engineering and emulation attacks.

8.4.4 Multiplexer-Based Arbiter PUF

The multiplexer-based arbiter PUF (MPUF) [135] adds non-linearity by incorporating a MUX output network, as shown in Fig. 8.5. A major advantage of the MPUF composition is that it can achieve higher reliability than a similar-sized XOR APUF instance while providing better security against modeling attacks. It has been shown in [135] that an (64, 3) MPUF is at least as robust as a 10-XOR APUF with a 64-bit challenge, and its reliability is as high as that of a 4-XOR APUF. The MPUF can be considered as a secure and reliable alternative for well-known XOR APUF in practice.

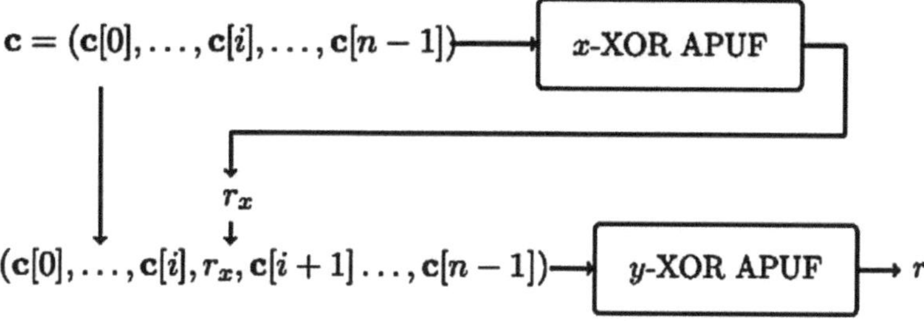

Fig. 8.6 A schematic representation of an (x,y) interpose PUF

8.4.5   The Interpose PUF

The Interpose PUF (iPUF) [119] aims at mitigating the known weaknesses of XOR Arbiter PUFs against modeling attacks. The iPUF architecture consists of a novel combination of two XOR Arbiter PUFs, as shown in Fig. 8.6. A challenge is fed to the first XOR-PUF block, where the computed response is fed as a challenge bit located in the middle of the second XOR-PUF block. The iPUF has been shown to increase resiliency against machine learning attacks while at the same time scaling better in terms of area overhead and reliability.

8.5   PUF-Based Authentication Protocols

Even a strong PUF primitive with ideal performance metrics and ML-based attack resilience would not be enough to establish secure authentication. To realize secure authentication, a tailored PUF-based authentication protocol must be employed. An authentication protocol would provide the needed security to strong PUFs that are prone to modeling attacks and, at the same time, offer security measures against threats such as man-in-the-middle, denial of service (DoS), and brute-force attacks. As Strong PUFs are vulnerable to modeling attacks, adversaries can produce a soft model of the PUF circuit, which can be used to correctly predict the response of any future PUF challenge.

Therefore, PUF-based authentication protocols tend to either employ resource-heavy PUF modules that are resilient against modeling attacks or utilize computationally heavy encryption or hashing functions [133]. However, both of these approaches would introduce a significant implementation overhead, making these protocols prohibitive for use in constrained devices. As an alternative, PUF-based authentication protocols can use simple pseudo-cryptographic algorithms that utilize TRNGs or PRNGs, which are less computationally heavy [108]. Such protocols can offer secure device authentication while exhibiting high resilience against modeling attacks. However, realizing lightweight and secure authentication

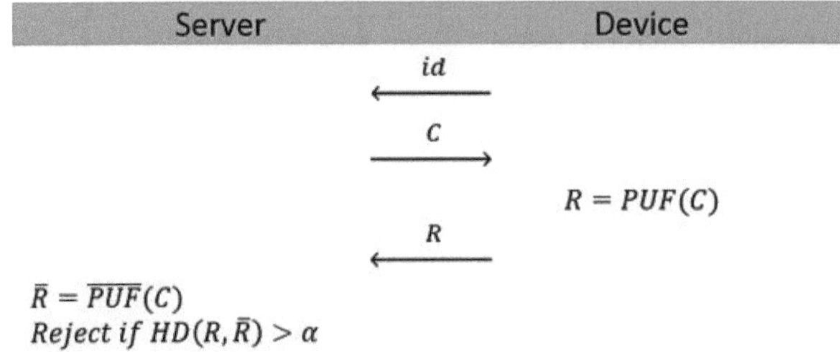

Fig. 8.7 A simplistic challenge-response exchange for device authentication

protocols capable of providing essential security features such as mutual device-server authentication, secret message exchange, device-to-device authentication, and resistance to DoS attacks is a very challenging task. That is why many of the proposed PUF-based protocols usually offer a combination of these security features but not all.

Figure 8.7 describes a challenge-response exchange scenario for authenticating a PUF device. A server with access to a soft model of the PUF would generate a set of challenges C and send it to the device. The device would then use its PUF circuit to generate a set of responses R for the received challenges, which are then sent back to the server. The server compares the device's responses with those generated from the soft model. If the received and generated responses match within a certain threshold α, the device is recognized as authentic. This authentication scenario is insecure, as an adversary can perform a modeling attack by collecting the exposed CRPs. Lightweight PUF-based authentication protocols utilize simple pseudo-cryptographic algorithms and avoid hash functions to satisfy the constrained devices' security needs.

8.6 Lightweight PUF-Based Authentication Protocols

While PUFs are gaining more popularity in security applications, only a handful of lightweight protocols that avoid cryptographic or hash functions have been introduced. However, most of the introduced lightweight protocols seem to be lacking in either their security or the features offered. The authors in [33] did a comprehensive survey on various PUF-based protocols. We highlight some of the most successful lightweight protocols along with more recently introduced work. Protocols that utilize hashing or other cryptographic primitives are not included.

8.6.1 Lockdown Protocol

Lockdown Protocol [166] limits the number of authentication attempts a device can perform. This ensures that a PUF would never reveal enough CRPs to allow an adversary to model the circuit. However, this also limits the number of authentications supported and opens vulnerabilities to denial of service (DoS) attacks that aim to exhaust the supported number of authentications. Overall, the protocol would be suitable for applications where DoS attacks might not be a concern, and a limited number of authentications would suffice for the device's lifetime. Furthermore, the lockdown does not offer a method to communicate authenticated secret messages, a feature that might be required depending on the application.

8.6.2 Slender Protocol

Slender Protocol [104] hides the PUF circuit's response in plain sight by concatenating the PUF responses with a random bit stream generated by a TRNG and then logically rotating the string by a random number to hide the index of the authentic response substring. In [9], it was shown that ES-CMA attacks could successfully produce an accurate model of the PUF circuit employed by Slender when observing a feasible number of authentications. This security vulnerability might be addressed by modifying Slender to add obfuscation to the exchange. However, Slender is unable to offer mutual authentication due to its computationally expensive verification process.

8.6.3 Zalivaka's Protocol

Zalivaka's Modeling-Resilient Arbiter PUF protocol [170] introduces a method to obfuscate the input challenges of an FPGA-based Arbiter PUF. The protocol utilizes a Multiple-Input Signature Register (MISR), a linear-feedback shift register, to implement a lossy compression algorithm on the raw challenges before feeding them to the PUF circuit. While resistant to modeling attacks, this approach requires strict synchronization of the seeds and the MISR between parties. This could be a critical vulnerability when Man-in-the-Middle attacks are possible, as altering intercepted messages can cause the parties to lose synchronization. Zalivaka's protocol cannot offer mutual authentication or secret message exchange.

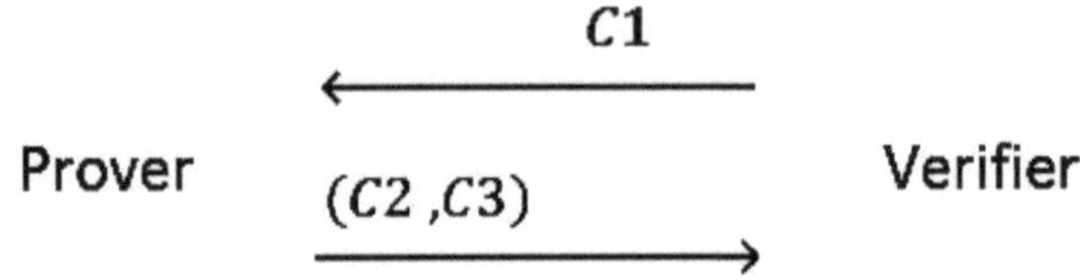

Fig. 8.8 The challenge-challenge exchange used for authentication

8.6.4 Obfuscated PUF-Based Authentication

Gu et al. [59] introduced a technique for obfuscated PUF-based authentication while utilizing delay-based lockdown to limit the overall number of CRPs that the adversary can collect in a fixed time. The technique uses fake responses when a delay threshold is exceeded to confuse attackers. This throughput-lockdown approach eliminates the vulnerability to denial of service attacks. However, Gu's protocol offers mutual authentication but does not offer secret message exchange.

8.6.5 Lightweight PUF-Based Authentication Protocol

A Lightweight PUF-based Authentication (LPA) protocol [71] uses secret pattern recognition to hide correlations between exchanged challenges and protect them against machine learning attacks such as CMA-ES, Deep Neural Networks (DNNs), and support vector machine (SVM). The protocol utilizes a challenge-challenge, as shown in Fig. 8.8, where the verifier challenge is generated by both the prover and verifier to protect against Man-in-the-Middle attacks. A desirable feature of the LPA protocol is the ability to offer both mutual authentication and authenticated secret message exchange, which are currently not offered by other lightweight PUF-based protocols.

8.6.6 Other Protocols and Summary

Other lightweight protocols that were investigated, like the Obfuscated Challenge-Response [50] and the Noise Bifurcation Protocol [168], showed good security. However, no testing has been done against CMA-ES-based attacks. Also, neither can offer mutual authentication. Depending on the application and available resources on the device, many of the proposed solutions could offer a suitable lightweight security solution for constrained IoT devices, with improved protocols capable of offering more security features that are expected to be realized from future research efforts.

8.7   A Reliable and Noise Resilient Strong PUF

In this chapter, we introduced a strong noise resilient PUF (NR-PUF) that employs a unique architecture, allowing it to extract more information from the circuit and drastically higher reliability. The NR-PUF offers better performance and avoids discarding any challenges. The NR-PUF achieves up to 98.7% reduction in error rate compared to the classic arbiter PUF while utilizing a very small implementation overhead, making it more desirable than other methods that employ soft response selection approaches [159, 162]. The NR-PUF also features a lower implementation area overhead when compared to other strong PUFs [62, 157, 165]. This dramatic increase in reliability, coupled with a very lightweight implementation, eliminates the need for expensive error correction circuitry and offers a truly lightweight strong PUF circuit suitable for constrained devices. A Mathematical Delay Model for the NR-PUF is provided and used for performance and security analyses. The reliability of the NR-PUF is verified through both software simulations and hardware implementations under various operating conditions. Furthermore, we show how the increase in reliability allows for better security scaling.

8.8   NR-PUF Circuit Architecture

The architecture of the NR-PUF utilizes four rows of N switching components, resulting in four symmetrically interconnected paths labeled A, B, C, and D, as shown in Fig. 8.9. Since there are four different delay paths, six different arbiters are required to compare the delay difference of all possible pairs. The output bits of each of the arbiters are referred to as AB, AC, AD, BC, BD, and CD (where AB = "1" when the signal propagation speed of path A is faster than that of B and "0" otherwise).

Switch components of the NR-PUF alternate between two different interconnections, Con-1 and Con-2, as shown in Fig. 8.9. By alternating between these two connection types, we preserve the routing symmetry between switch components. We have experimented with several connection designs, and the 2-type scheme adopted showed the best symmetry and a highly unbiased output for the PUF circuit, a quality that is essential to the PUF circuit security. The design choice of the interconnections is of critical importance as it has a direct impact on the bias and, subsequently, the randomness of the PUF's responses. By utilizing two alternating connection types, we were able to produce a highly symmetrical circuit while also allowing all four delay elements to be part of any of the four racing paths and to fairly compete with one another, resulting in unbiased outputs. This was confirmed through our implementation and testing, which are presented in a later section.

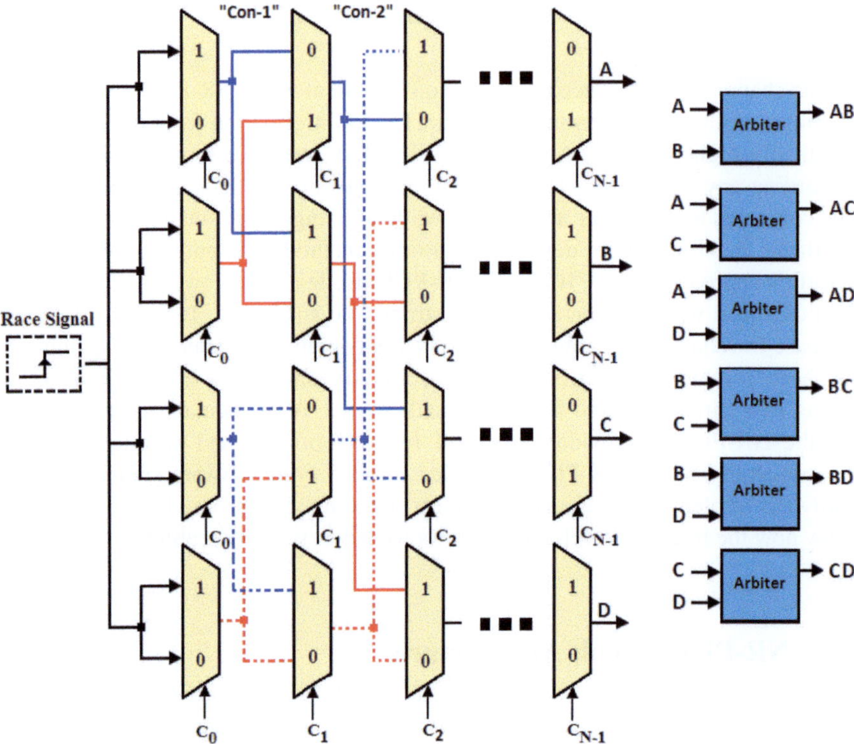

Fig. 8.9 Noise resilient PUF circuit architecture

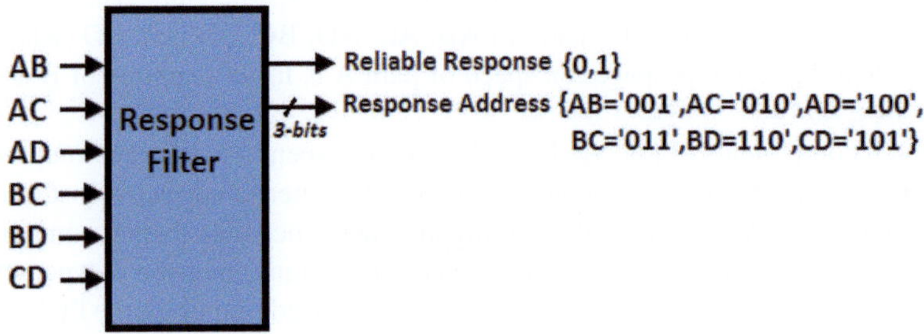

Fig. 8.10 Logic block of the utilized response selection circuitry

8.9   NR-PUF's Reliable Response Selection

A response selection circuitry is utilized to select highly reliable responses with high delay differences, as shown in Fig. 8.10. A total of six arbiters are used to compare the delays among all four symmetrical paths of the circuit. The logic circuit would analyze the outputs of each of the six arbiters and produce the order in which the signals A, B, C, and D have arrived. The selection logic is a truth table described

Arbiters' Outputs	Sorted State	HiRel , Addr	Arbiters' Outputs	Sorted State	HiRel , Addr	Arbiters' Outputs	Sorted State	HiRel , Addr
111111	ABCD	1 , 100	001111	BCAD	1 , 110	100001	CDAB	0 , 011
111110	ABDC	1 , 010	000111	BCDA	0 , 001	000001	CDBA	0 , 010
111011	ACBD	1 , 100	010110	BDAC	1 , 011	110100	DABC	0 , 101
111001	ACDB	1 , 001	000110	BDCA	0 , 001	110000	DACB	0 , 110
111100	ADBC	1 , 010	101011	CABD	1 , 101	010100	DBAC	0 , 101
111000	ADCB	1 , 001	101001	CADB	0 , 011	000100	DBCA	0 , 100
011111	BACD	1 , 110	001011	CBAD	1 , 101	100000	DCAB	0 , 110
011110	BADC	1 , 011	000011	CBDA	0 , 010	000000	DCBA	0 , 100

Fig. 8.11 The twenty-four theoretically possible responses and their corresponding sorted states (e.g., if the response of arbiter AB=1, AC=1, AD=1, BC=1, BD=1, and CD=1, then the sorted state is ABCD and "HiRel, Addr" = "1100")

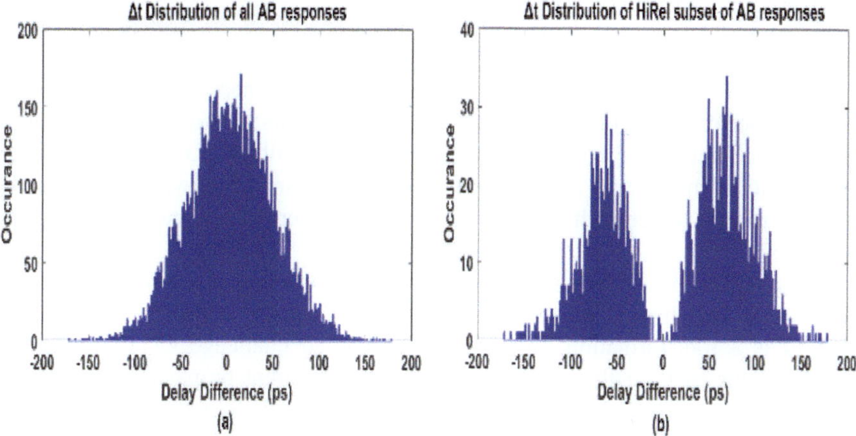

Fig. 8.12 Delay distribution of the NR-PUF's responses when including: (**a**) all responses of arbiter AB, (**b**) responses with A and B are two positions apart (e.g., ACDB)

in Fig. 8.11. Depending on the outputs of the six utilized arbiters, the output of the race with the highest delay difference estimate $|\Delta t|$ would be selected as a Highly Reliable response bit (HiRel) of the NR-PUF. A 3-bit response address (Addr) would also be used to identify which of the six races/arbiters has been selected. In the selection circuitry, the race signals of each path are in descending order, from fastest to slowest (e.g., "BDCA" where B is the fastest and A is the slowest of the four racing signals). Sorted signals will belong to a pool of 24 different states (i.e., race results), as shown in Fig. 8.11. By choosing the arbiter's response with the highest delay difference $|\Delta t|$ value, we can guarantee highly stable PUF responses.

A mathematical model was used to initially test the effectiveness of the proposed architecture in increasing the average delay difference $|\Delta t|$ at the output. Figure 8.12 shows the delay difference $|\Delta t|$ of 20k random CRPs when including all the responses of Arbiter "AB" (Fig. 8.12a), versus utilizing the filtering scheme to only include Highly Reliable (HiRel) responses (Fig. 8.12b). The result shows clearly how the selected HiRel response (Fig. 8.12b) has a distribution that avoids the region

$|\Delta t| = 0$, while unfiltered responses (Fig. 8.12a) are heavily centered on $|\Delta t| = 0$, making them more vulnerable to noise.

8.10 NR-PUF's Mathematical Delay Model

The NR-PUF consists of k stages of delay elements; most commonly, these delay elements would be multiplexers. Four signals will race simultaneously through each of the k stages. The exact path of the signals is controlled by an applied challenge vector $\mathbf{C} = c_1, c_2, ..., c_k$, where c_k belongs to 0, 1. Finally, six different arbiters, such as SR-latches or D-Flip-Flops (DFFs), will compare the propagation delay of the paths.

The conventional way to describe the functionality of an Arbiter PUF is via an additive linear delay model [105, 131], which would also be accurate for our proposed NR-PUF. Such a mathematical delay model is necessary to evaluate the different performance metrics of a PUF and to evaluate its security against modeling attacks that employ supervised machine learning algorithms [131]. Additionally, a mathematical model will also help in efficiently storing a software model of the PUF on trusted servers, which will be utilized in authentication and key generation schemes.

In an APUF, the final delay difference Δ of the existing two paths is expressed as $\Delta = \mathbf{W}^T \mathbf{\Theta}$, where $\mathbf{W}$ represents the delay parameters of an APUF's components and $\mathbf{\Theta}$ is a transformation function of the applied challenge $\mathbf{C}$ [131]. For the NR-PUF, for each challenge $\mathbf{C}$, there exist four different propagation paths and, consequently, six different delay difference comparisons (i.e., arbiters AB, AC, AD, BC, BD, and CD) that are to be computed. We describe the linear additive delay model of an NR-AUF as follows. Let δ_n^{ij} represent the delay difference between the ith and the jth delay elements at stage n, as shown in Fig. 8.13, and then:

$$\delta_n^{ij} = \delta_n^{i} - \delta_n^{j}, \ \text{for all } i, j = 1, \ldots, \ 8 \text{ and } n = 1, \ldots, k \text{ where } i \neq j. \tag{8.3}$$

And $\vec{K_0}$, $\vec{K_1}$, $\vec{K_2}$, and $\vec{K_3}$ represent the different transformation vectors of the input challenge $\vec{C}$, where:

$$\vec{K}_0^{\,j} = \left(1 - 2c_j\right), \ \text{for } j = 1, \ldots, k \tag{8.4}$$

$$\vec{K}_1^{\,j} = \left(\vec{K}_1^{\,1}, \vec{K}_1^{\,2} \ldots, \vec{K}_1^{\,k}, 1\right), \text{where } \vec{K}_1^{\,j} = \prod_{i=1}^{k} \vec{K}_0^{\,j} \text{ for } j = 1, \ldots, k. \tag{8.5}$$

$$\vec{K}_2 = \left(\vec{K}_2^{\,2}, \vec{K}_2^{\,3}, \ldots, \vec{K}_2^{\,k-1}, 1\right), \text{where } \vec{K}_2^{\,j} = \vec{K}_2^{\,j+1} = \prod_{i=1}^{\frac{k}{2}} \vec{K}_0^{\,2i},$$

$$\text{for } j = 1, 3, 5 \ldots, \ k-1. \tag{8.6}$$

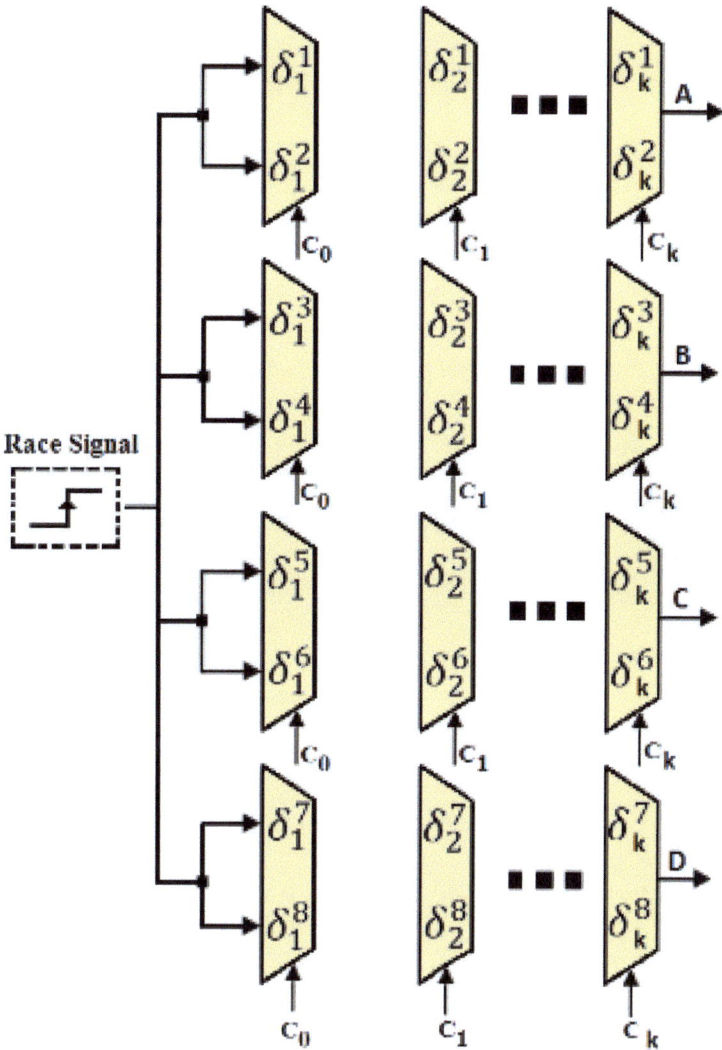

Fig. 8.13 Illustration of the delay elements applied in the NR-PUF mathematical model

$$\vec{K}_3 = \left(\vec{K}_3^{\,1}, \vec{K}_3^{\,2} \ldots, \vec{K}_3^{\,k-2}, 1, 1 \right), \text{ where } \vec{K}_3^{\,j} = \vec{K}_3^{\,j+1} = \prod_{i=2}^{\frac{k}{2}} \vec{K}_0^{\,2i-1}$$

$$\text{for } j = 1, 3, 5 \ldots, k-3. \tag{8.7}$$

Let k be the size of the PUF (e.g., the number of stages/multiplexers in each row).
The partial weight parameters' vectors $\vec{w}_1$, $\vec{w}_2$, $\vec{w}_3$, $\vec{w}_4$, and $\vec{w}_5$ are defined as

follows:

$$\vec{w}_1 = \left(w_1^1,\, w_1^2 \ldots,\, w_1^k,\, w_1^{k+1} \right),$$ (8.8)

where : $w_1^1 = \delta_1^{12} - \delta_1^{56} - \delta_1^{34} + \delta_1^{78}$, and $w_1^i = \delta_1^{12} - \delta_1^{56} - \delta_1^{34} + \delta_1^{78} + \delta_{i-1}^{12}$

$$+ \delta_{i-1}^{56} - \delta_{i-1}^{34} - \delta_{i-1}^{78}$$

for all $i = 2, \ldots, k$, and $w_1^{k+1} = \delta_k^{12} + \delta_k^{56} - \delta_k^4 - \delta_k^{78}$.

$$\vec{w}_2 = \left(w_2^1,\, w_2^2 \ldots,\, w_2^k \right),$$ (8.9)

where : $w_2^i = \delta_i^{12} - \delta_i^{56} + \delta_i^{34} - \delta_i^{78}$ for all $i = 1, \ldots, k$.

$$\vec{w}_3 = \left(w_3^1,\, w_3^2 \ldots,\, w_3^k \right),$$ (8.10)

where : $w_3^i = \delta_i^{12} + \delta_i^{56} + \delta_i^{34} + \delta_i^{78}$ for all $i = 1, \ldots, k$.

$$\vec{w}_4 = \left(w_4^1,\, w_4^2 \ldots,\, w_4^k \right),$$ (8.11)

where $w_4^i = \delta_i^{13} - \delta_i^{57} + \delta_i^{24} - \delta_i^{68}$ for all $i = 1, \ldots, k$

$$\vec{w}_5 = \left(w_5^1,\, w_5^2 \ldots,\, w_5^k \right),$$ (8.12)

where $w_5^i = \delta_i^{13} + \delta_i^{56} + \delta_i^{24} + \delta_i^{68}$ for all $i = 1, \ldots, k$.

The outputs of arbiter AB, AC, AD, BC, BD, and CD of the NR-PUF are thus determined by the sign of the final delay difference, where a value of "-1" or "1" represents an arbiter's response of "0" or "1," respectively. The Arbiter's responses are defined as follows (where $\circ$ represents an element-wise multiplication of two vectors):

$$AB = sgn\left[\vec{K}_1 \vec{w}_1{}^T + \left(\vec{K}_2 \circ \vec{K}_0 \right) \vec{w}_2{}^T + \vec{K}_2 \vec{w}_3{}^T \right]$$ (8.13)

$$AC = sgn\left[\vec{K}_1 \vec{w}_1{}^T + \left(\vec{K}_3 \circ \vec{K}_0 \right) \vec{w}_4{}^T + \vec{K}_3 \vec{w}_5{}^T \right]$$ (8.14)

$$AD = sgn\left[\begin{array}{c} \left(\vec{K}_2 \circ \vec{K}_0 \right) \vec{w}_2{}^T + \vec{K}_2 \vec{w}_3{}^T \\ + \left(\vec{K}_3 \circ \vec{K}_0 \right) \vec{w}_4{}^T + \vec{K}_3 \vec{w}_5{}^T \end{array} \right]$$ (8.15)

$$BC = sgn \begin{bmatrix} -\left(\vec{K}_2 \circ \vec{K}_0\right)\vec{w}_2^T - \vec{K}_2\vec{w}_3^T \\ +\left(\vec{K}_3 \circ \vec{K}_0\right)\vec{w}_4^T + \vec{K}_3\vec{w}_5^T \end{bmatrix} \tag{8.16}$$

$$BD = sgn \left[-\vec{K}_1\vec{w}_1^T + \left(\vec{K}_2 \circ \vec{K}_0\right)\vec{w}_2^T + \vec{K}_2\vec{w}_3^T\right] \tag{8.17}$$

$$CD = sgn \left[-\vec{K}_1\vec{w}_1^T + \left(\vec{K}_3 \circ \vec{K}_0\right)\vec{w}_4^T + \vec{K}_3\vec{w}_5^T\right]. \tag{8.18}$$

In our model-based experiments, we use the defined NR-PUF delay model to evaluate its performance. The APUF delay model is also used to compare its performance with our proposed NR-PUF. Delay elements δ_n^i in our mathematical delay model would belong to a Gaussian distribution of mean $\mu = 0.5$ ns and standard deviation $\sigma = 4$ ps, representing the mean and variance of 65 nm CMOS technology [139]. Random noise is added to the computed delay of each path, following a Gaussian distribution of mean $\mu = 0$ and standard deviation σ varying between 0.3 and 11, to simulate the expected BER of a basic Arbiter PUF at varying operating conditions that were observed in [60, 162].

8.11 NR-PUF's Performance

8.11.1 Reliability

The NR-PUF achieves a dramatic increase in reliability when compared to the Arbiter PUF circuit. A PUF reliability is a measure of the stability of the PUF responses when the same challenge is evaluated under varying operating conditions. The BER is evaluated at varying induced noise levels belonging to a normal distribution $N(\mu, \sigma^2)$ with $\mu = 0$ and $\sigma = \alpha$. The choice of α is meant to replicate the observed BER of an APUF when implemented on 65 nm CMOS technology [60, 162] under varying operating conditions, where a native $BER \simeq 4\%$ and a worst-case $BER \simeq 10\%$ were reported.

From Fig. 8.14, we can see that when utilizing the introduced NR-PUF, the BER is reduced considerably. In a worst-case scenario when the PUF is simulated to be operating in extremely noisy conditions, an APUF has a BER of 10.39%, while the NR-PUF has an average BER of 1.06%. An even more drastic increase in performance is observed when the PUF is operating at nominal operating conditions. While an APUF has an average BER of 4.04%, the introduced NR-PUF can achieve a BER as low as 0.052%, a 98.7% reduction in error rate. The NR-PUF can always guarantee the availability of a HiRel response with a higher mean value $|\Delta t|$ than that of an APUF's response. As a result, its reliability is always superior to that of an APUF. Compared to the APUF's BER, our NR-PUF can thus achieve a BER

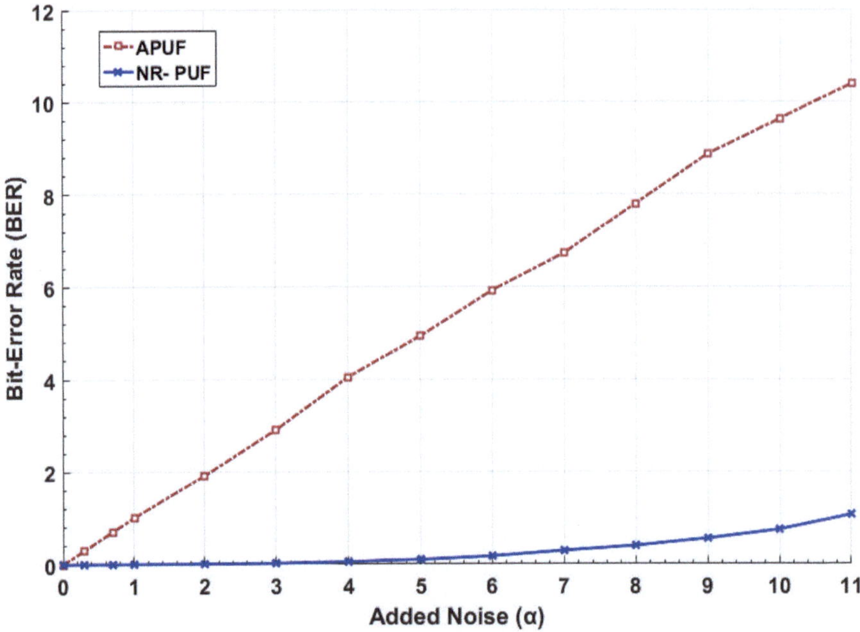

Fig. 8.14 The BER for a 64-bit APUF and a 64-bit NR-PUF at different noise levels

reduction that varies between 89.8% and up to 98.7% depending on the operating condition.

8.11.2 Resilience to Aging

Aging is another problem that directly affects the repeatability of the PUF's responses. NBTI/PBTI (negative/positive bias temperature instability) and HCI (hot-carrier injection) are two leading factors in the performance degradation of digital circuits over time [120]. The aging effect on delay-based PUFs and methods to mitigate it has been a subject of interest in many literature studies [74, 149]. The reliability of Arbiter PUFs has been shown to degrade by about 2.41% after ten years [115], which is mostly due to the impact of aging on the utilized arbiters at the end of the APUF's delay chains [74]. Aging at the arbiter would introduce a bias toward "0" or "1" when comparing between two delay paths, leading to undesirable bit-flips at the arbiter's output. The effect of such induced bias on the NR-PUF's reliability is investigated. A bias value of $|\Delta t^a|$ is added to the computed delay difference of the two paths at each arbiter, to emulate the effect of aging at the arbiter and examine how it affects the reliability of the PUF's responses.

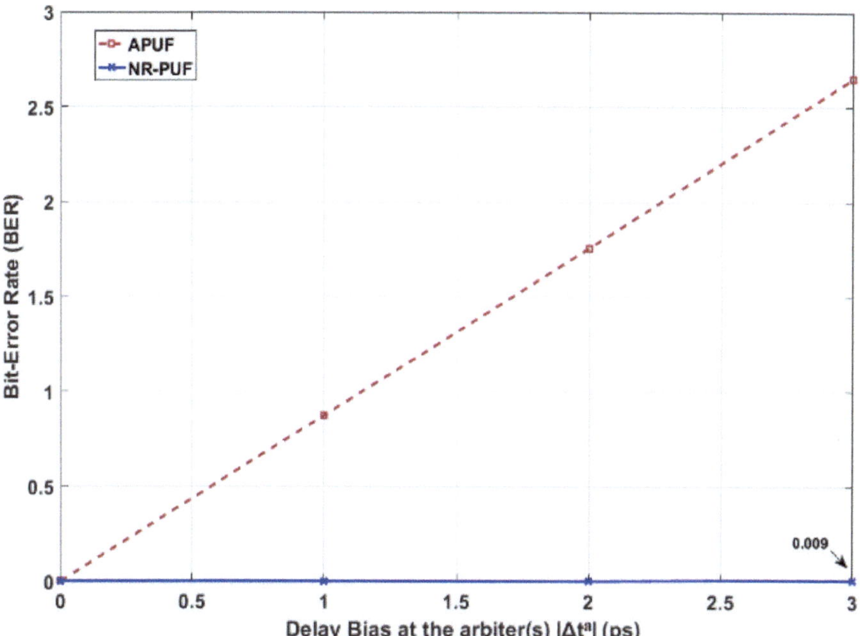

Fig. 8.15 The BER of the APUF and the NR-PUF when introducing a delay bias $|\Delta t^a|$ at their arbiters

Table 8.1 Software model performance results of 64-bit APUF and NR-PUF

	NR-PUF	APUF
Randomness (μ, σ)	(46.47, 2.6)	(47.22, 2.1)
Uniqueness (μ, σ)	(46.45, 2.68)	(46.81, 2.41)
Mean BER $(\alpha = 4)$	0.052%,0.09%^{w}	4.04%,4.79%^{w}
Mean BER $(\alpha = 11)$	1.06%, 1.2%^{w}	10.39%, 12.18%^{w}

^{w} Worst BER observed

Figure 8.15 shows the BER of both the NR-PUF and the APUF under varying induced bias values $|\Delta t^a|$. Results show that aging at the NR-PUF's arbiters has little effect on the reliability of its responses, while the APUF's reliability is very sensitive to any introduced bias or aging at its arbiter. This is expected since we know that most of the APUF's responses have a small delay difference $|\Delta t|$, while few of the NR-PUF's responses are as such, which has been shown in Fig 8.12 (b). Table 8.1 summarizes the measured quality metrics of both the NR-PUF and the basic APUF. The proposed NR-PUF has shown to offer superior reliability compared to that of an APUF, while maintaining the same high randomness and uniqueness qualities.

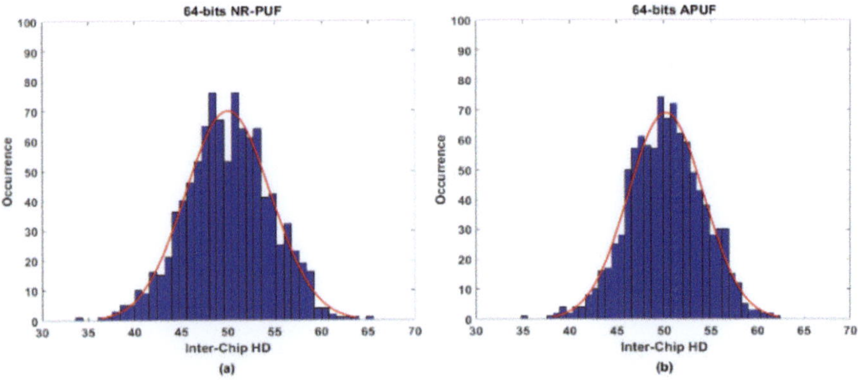

Fig. 8.16 Inter-chip comparison of a 64-bit NR-PUF and a 64-bit APUF

8.11.3 Randomness

The NR-PUF would have four outputs: a 1-bit HiRel response value of "0" or "1" and a 3-bit address Addr value pointing to one of the six arbiters that have been selected. A defining feature of a PUF circuit is the unpredictability of its responses, which is reflected by its randomness. The measured average randomness H of the NR-PUF's HiRel responses and the APUF's responses are 46.47 and 47.22, respectively. This is a very high randomness quality that approaches the ideal value of $H = 50$.

8.11.4 Uniqueness

The uniqueness between PUF devices is measured by calculating the Inter-Hamming Distance (Inter-HD) of its responses with other PUF instances. In our experiment, 500 pairs of PUF instances are initialized, and the Inter-HD of 10,000 CRPs is measured. Figure 8.16a shows the histogram of the measured Inter-HD of 500 NR-PUF pairs, while Fig. 8.16b shows the Inter-HD of 500 APUF pairs. The APUF has shown to have an average uniqueness of Inter-HD = 46.8, and the NR-PUF has shown to have a very similar average Inter-HD. This confirms that the introduced NR-PUF has a uniqueness value that is very close to the optimal value of 50.

8.12　NR-PUF's Circuit Area Overhead

The NR-PUF consists of k stages of delay elements; most commonly, these delay elements are multiplexers. Four signals will race simultaneously through each of the k stages. The exact path of the signals is controlled by an applied challenge vector

Table 8.2 Area comparison of the NR-PUF and APUF, when considering BCH error correction codes

	NR-PUF	APUF
PUF area cost (GE)	760	380
Response selector (GE)	141	NA
Failure rate for a 63-bit code word	3.2%	92.56%
BCH decoder (GE) [91]	NA	5580 , BCH (63,36,5)
Code-word failure rate after BCH	BCH 3.2%	4.16%
Total area (GE)	901	5960

$\mathbf{C} = c_1, c_2, ..., c_k$, where c_k belongs to 0, 1. Finally, six different arbiters, such as SR-latches or D-Flip-Flops (DFFs), will compare the propagation delay of the paths. In order to highlight the advantage of using the NR-PUF over utilizing error correction codes such as BCH (Bose–Chaudhuri–Hocquenghem) error correction codes, we compare the area overhead associated with introduced NR-PUF vs. that introduced when implementing the BCH error correction circuit. The results showed that by utilizing the NR-PUF, we can drastically reduce the implementation area of a system that is utilizing an error correction code.

To ensure an impartial comparison, both the NR-PUF and the APUF were implemented in 45 nm CMOS technology using the FreePDK45 digital library [143]. Synthesis results were generated by Synopsys Design Vision. The area requirement of each design is reported in gate equivalent (GE) units. The area footprint of a 64-bit APUF is equivalent to 380 GE, while a 64-bit NR-PUF would require 760 GE in addition to a 141 GE for the Response Selection Logic (RSL) for a total of 901 GE. It is important to note that the area requirement of the RSL is fixed, regardless of the implemented NR-PUF's size (i.e., bit length). As the APUF has the smallest implementation area cost among known Strong PUFs to date, the area overhead of our proposed PUF is very reasonable, especially when considering the reliability gain achieved. Table 8.2 compares the total area requirement for both the APUF and the NR-PUF, if both PUF designs were to achieve similar reliability for a 63-bit code word. The choice of the BCH code is based on the equation provided in [126], which is defined as $P_{ecc} = \sum_{i=b+1}^{n} P_{err}^i (1 - P_{err})^{n-i}$, where P_{err} is the nominal BER of the utilized PUF, n is the code length, and b is the number of correctable bit errors. An APUF employed with a BCH(63,36,5) would still have a higher code-word failure rate than our introduced NR-APUF with no error correction.

Utilizing the NR-PUF would drastically reduce the total implementation area. At nominal operating conditions, the implementation area overhead is reduced by a minimum of 84.8%. However, as aging and noise increase the BER, which is expected in a real operating environment, the NR-PUF would have an even greater advantage against circuits that rely solely on error correction codes that have their implementation area grow rather exponentially when attempting to correct a larger number of errors.

8.13 NR-PUF's Security

Using the mathematical model of both the APUF and the NR-PUF, the security against modeling attacks of both the original APUF and our introduced NR-PUF is evaluated. A supervised machine learning (ML) framework using Linear Regression (LR) with RProp [126] gradient descent was employed to model both PUFs. LR with RProp has been chosen because it has shown teruhrmair2013puf to be the most efficient algorithm when modeling APUFs. A test set of 10,000 randomly generated CRPs is used to evaluate the prediction accuracy of the ML algorithm at different training set sizes. The process of collecting CRPs when training the NR-PUF slightly differs from that of the traditional APUF, since only the HiRel Response of a corresponding challenge can be observed. However, an adversary would be capable of deriving a total of five output values of the NR-PUF's responses based on the observed HiRel response bit and its corresponding arbiter's 3-bit address, contrary to all six outputs. For example, with a HiRel response bit value of "1" and Addr = "001" pointing to arbiter AB, we know that the response' sorted state is either ACDB or ADCB with only the response of arbiter CD unknown.

Figure 8.17 shows the prediction accuracy of the modeled PUF when different numbers of CRPs (NCR) are used to train each model. An adversary would require a similar number of CRPs to accurately predict all six responses of the NR-PUF as when predicting the response of an APUF. Such results were expected, knowing that both PUF has similar linearity and can be described in a linear additive delay

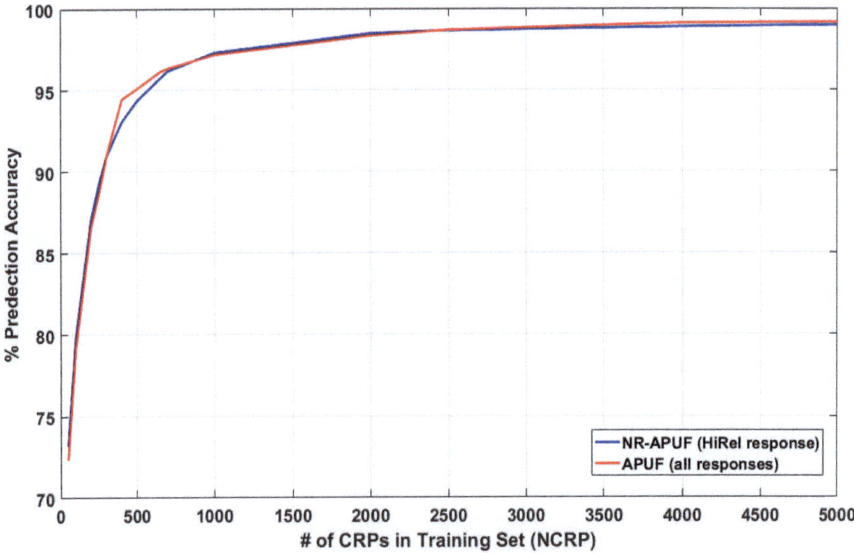

Fig. 8.17 The accuracy of the modeled PUF using supervised machine learning in relation with the number of observed CRPs used in the training set

model. However, because of the high stability of the NR-PUF's responses, allows it to scale better in performance and security when used in controlled strong PUF designs such as XOR-ing multiple PUF instances, where the reliability of the PUF's responses is exponentially diminished with the number of PUFs XOR-ed together.

8.14   NR-PUF's Hardware Verification

The reliability performances of the NR-PUF and the APUF were tested on two different FPGA setups: (i) A 64-bit APUF along with a 64-bit NR-PUF were implemented on an Artix-7 FPGA using built-in mux modules in the FPGA as delay elements; (ii) In another experiment, a 64-bit APUF and a 64-bit NR-PUF are synthesized using the delay values of Virginia Tech's public FPGA dataset [103] of five different RO PUFs. Both experiments were consistent with the estimates produced by the mathematical model, thereby achieving a 99.86% reduction in BER at nominal operating.

8.14.1   FPGA Implementation

A 64-bit NR-PUF along with a 64-bit APUF were implemented on a CW305 Artix-7 FPGA board to test the reliability of the responses on varying voltages, as well as on two Arty Artix-7 FPGA boards to test their uniqueness. Experiment results are evaluated using a set of 20k challenges randomly generated using MATLAB [111]. Responses of each challenge are repeatedly evaluated 7 times at $T = 35\,°C$ and V $= 1.2\,V$, where a golden set of CRPs is attained based on majority voting. The same challenges were also evaluated at varying voltages ranging from $0.8\,V$ to $1.05\,V$ with a $0.5\,V$ increment interval. This experiment was repeated across 10 different instances of NR-PUFs and APUFs.

The average BER of the NR-PUF is compared with the average BER of an APUF at varying operating voltages, as shown in Fig. 8.18. The BER of an APUF drastically increases when operating at varying voltage levels than optimal. An APUF achieves a nominal BER of 0.3% when operating at $1.2\,V$ and a worst BER of 4.25% when operating at $0.8\,V$, while the NR-PUF maintained a low BER of 0.0015% and 0.069% when operating at $1\,V$ and $0.8\,V$, respectively. The NR-PUF's achieved BER reduction compared to that of an APUF agrees with our mathematical model-based simulations.

8.14.2   Synthesized RO Structured Implementation

The utilized Virginia Tech's public dataset includes five different RO PUFs implemented across five Spartan3E S500 FPGA boards. Each RO PUF implementation

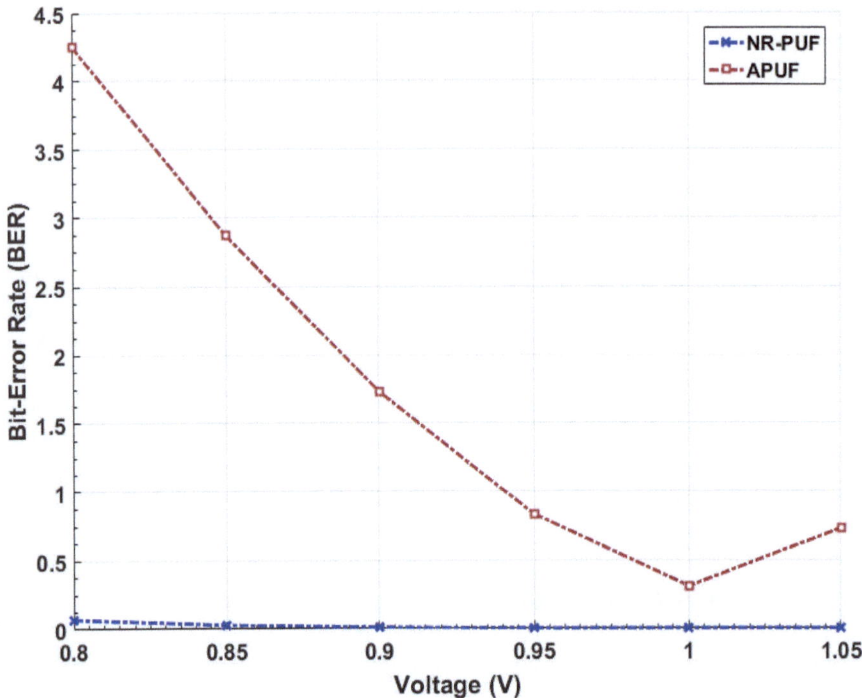

Fig. 8.18 BER of both APUF and NR-PUF FPGA implementations, observed at varying operating voltages

consists of 512 ring oscillators (ROs), where each of the RO's frequency and count is measured 100 times at core supply voltages of 0.98 V, 1.08 V, 1.20 V, 1.32 V, 1.44 V and at temperatures of 35 °C, 45 °C, 55 °C, 65 °C. Variable voltage measurements were taken at room temperature $T = 35\,°C$, while variable temperature measurements were taken at a fixed voltage $V = 1.2\,$V. Measured RO counts were used as delay elements in Eq. 8.1 and Eqs. (8.13–8.18) to realize a 64-bit APUF and a 64-bit NR-PUF. The final delay of each of the four paths is measured by calculating the sum of 64 RO counts that are decided based on the input challenge vector **C**. As a result, we were able to obtain five synthesized NR-PUF instances and ten synthesized APUFs, where the performance of each PUF is evaluated using a set of 100,000 randomly generated challenges. Figure 8.19a shows the average BER of an NR-PUF compared to that of an APUF at varying operating temperatures, while Fig. 8.19b shows the average BER when operating at varying operating voltages. In Fig. 8.19a, when the PUF is operating under varying temperatures, the average BER of an APUF ranges between 1.07% and 1.95%, while the BER of an NR-PUF between 0.0015% and 0.0075%, showing that the NR-PUF is hardly affected by the imposed by temperature variations. Figure 8.19b shows the average BER of both PUFs when operating at varying voltage variations. The APUF BER varies

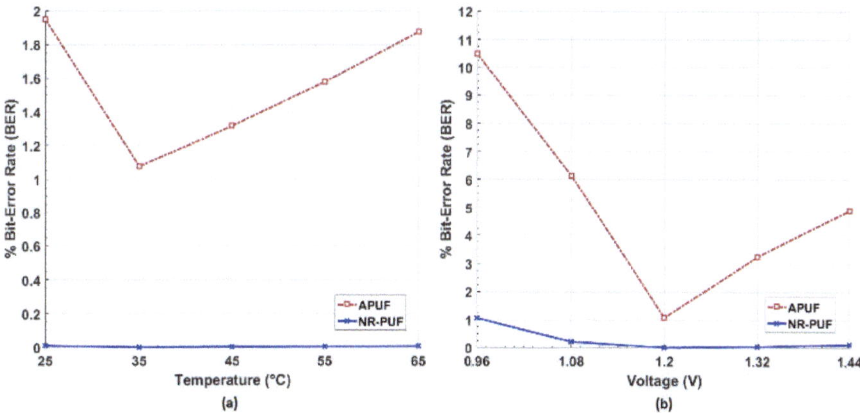

Fig. 8.19 BER of the synthesized 64-bit APUF and NR-PUF, observed at varying (**a**) temperature and (**b**) voltage conditions

between 1.07% and 10.47%, while NR-PUF shows a much lower BER varying between 0.0015% and 1.05%. Based on these results, we were able to achieve a minimum of 89.97% in BER reduction, which is identical to the expected value of 89.8% estimated in our mathematical model's software simulations.

Furthermore, we tested the effect of applying a repetition code of size t. In a t repetition code, a challenge C is computed t-times. If the response is consistently the same for all t computations (i.e., consistently all zeros or ones at the same initially selected arbiter Addr), the challenge is deemed reliable and used for authentication. Otherwise, it is discarded. By applying a repletion code of $t = 30$, we can achieve a nominal BER of 0%, and the worst BER of 0.606% when it was originally at 1.05% (i.e., operating at $V = 0.96V$), while the reliability we gain when applying a $t > 30$ will be relatively minuscule in comparison.

8.15   NR-PUF's Summarized Features

Increasing a strong PUF's reliability is essential for establishing secure, lightweight PUF protocols that can be used in constrained devices. This work introduces a new Noise Resilient PUF design architecture exhibiting high uniqueness and randomness qualities and capable of generating stable responses that are resistant to aging as well as temperature and voltage variations. The NR-PUF features a low BER of 0.052% when operating at nominal conditions, which is equivalent to a 98.7% BER reduction compared to an APUF. The NR-PUF can further benefit from soft error correction approaches such as temporal majority voting and repetition codes. A worst-case BER of 0.606% can be achieved with a $t = 30$ repetition code. With such a highly reliable PUF, we can avoid the need for costly ECC circuitry and

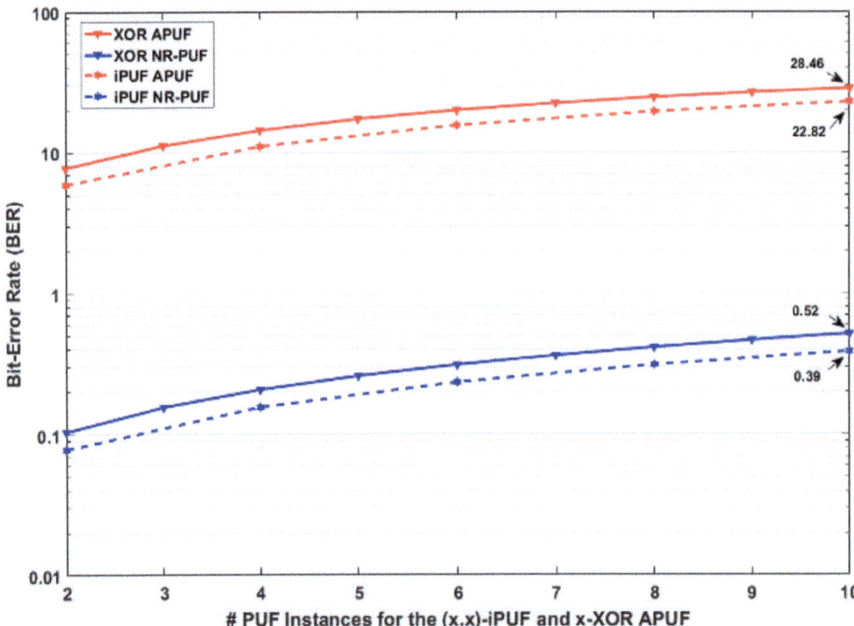

Fig. 8.20 Expected BER when scaling the security of the APUF and NR-PUF through traditional XOR-ing and the (x/2,x/2)-iPUF, when both designs are operating at nominal conditions

allow for better security scaling. Figure 8.20 shows how our proposed NR-PUF's reliability would scale when used with XOR-ing or the Interpose PUF (iPUF) [119] approach, in comparison to that of an APUF. The iPUF is a variation of XOR-ing and Feed-Forwarding architecture that aims to provide better scalability than just XOR-ing. The NR-PUF would allow for a higher number of instances to be utilized, while still maintaining acceptable functionality. A (5,5)-iPUF utilizing the NR-PUF design would have a nominal BER of 0.39%, as an alternative to a BER of 22.82% when using the basic APUF design. The results show that by enhancing the PUF's response reliability, we were also able to provide better security against modeling attacks without sacrificing functionality. The NR-PUF's reliability was confirmed using two different FPGA setups and showed consistent results with those based on the software model, by providing the same expected BER reduction when compared to an APUF.

Chapter 9
The Right Fit to Your Device and Application

Choosing the appropriate PUF for a specific device and application is crucial to maximizing both security and performance. This chapter provides a comprehensive guide on selecting the right PUF technology tailored to different operational requirements and constraints. By examining various PUF types and their respective strengths and weaknesses, we offer insights into how to match PUF characteristics—such as area, power consumption, response time, and reliability—to the unique demands of diverse applications. Whether for lightweight IoT devices, high-security authentication systems, or large-scale data centers, understanding the alignment between PUF capabilities and application needs is essential for deploying effective and efficient security solutions.

9.1 PUF Security Efficiency and Scaling

The security of different PUFs designs against modeling attacks has been evaluated in multiple research efforts. However, a more accurate evaluation that incorporates the area footprints of the different PUFs is missing from the literature. Such evaluation is rather essential in order to determine the most suitable PUF design for the different constrained devices. To perform this evaluation, we implemented several strong delay-based PUFs in 45 nm technology. The results showed that when factoring in the PUFs implementation area costs, these PUFs compare differently as some enhanced PUF designs turned out to have inferior efficiency compared to their simpler counterpart. We recognized the most effective design elements in the implemented PUFs. Based on the efficiency figures obtained, we recommended the optimal PUF design for the different security schemes used in constrained IoT devices.

K. Khalil et al., *Lightweight Hardware Security and Physically Unclonable Functions*, https://doi.org/10.1007/978-3-031-76328-1_9

9.2   Resistance to Modeling Attacks

The delay-based Arbiter PUF and its variants were shown to be vulnerable to modeling attacks in [132, 152], and [48]. Such attacks utilize machine learning methods to correctly predict the output of new challenges by analyzing extracted CRPs (Challenge-Response Pairs). These CRPs can be extracted by physically probing the circuit itself, by continuously listening and recording authentication messages received and sent by the device, or by impersonating a legitimate authenticator and polling the device repetitively. The security of a given PUF, in regards to resisting modeling attacks, can be parameterized by two values [132]:

1. NCR: the number of CRPs that need to be collected in order to successfully model the PUF with 99% predictability
2. TTM: the time required to successfully model a PUF, given that enough CRPs are collected

For the PUF to be secure, it has to resist modeling attacks either by requiring a large number of CRPs that are infeasible to collect (a very large NCR) or by requiring an infeasible amount of time to produce a model of the PUF (a very large TTM).

9.3   PUF Security Schemes in IoT Applications

In this section, we classify the PUF security schemes based on an adversary's ability to collect exchanged CRPs. We recognize three different accessibility levels for the CRPs: hidden, limited, and fully accessible. We mention the possible applications and IoT devices that might employ each.

9.3.1   CRP Hidden

In such schemes, the transmitted challenges and responses are completely hidden from an adversary. This is the case for many Controlled PUFs [53] (CPUFs), where the transmitted challenge and response are transformed through cryptographic functions or pseudo-cryptographic functions. In these schemes, the device should also be secured against probing attacks. This can be done either by denying physical access or by utilizing probe-resistant technology.

For reliable security, it is essential to use proven cryptographic functions at the cost of increased area requirements. Such schemes are best suited for personal low-power mobile devices that are under their owner's supervision. These can include wearable computing or medical devices that have essential security needs and are critically powered. For these devices, a PUF's NCR and TTM values are rather irrelevant as we assume it is impossible to model the PUF due to the lack of access to the CRPs. As such, we recommend PUFs with the lowest area implementation requirement, as the control logic used to hide the CRP exchanges will add a significant area overhead.

9.3.2   CRP Limited

In this scheme, the CRPs can be collected in a limited fashion. The attacker might only be able to collect the CRPs transmitted and requested by trusted parties. Another possibility is having the PUF polling rate artificially slowed to ensure that an adversary would not be able to poll the device at a sufficient speed. Probing attacks would still be a problem, and the device should be secured against them in a similar fashion to the PUFs used in the CRP Hidden scheme. This PUF security scheme does not require cryptographic control functions, and as such, it is suitable for devices with stricter area/power limitations. These devices can be ultra-low-power devices such as passive RFID tags used for personal identification like in passports or ID cards. The security of PUFs implemented in such schemes would be entirely dependent on the achieved NCR value.

9.3.3   CRP Fully Accessible

In this scheme, the required number of CRPs to successfully model a PUF circuit can be collected by an adversary either by polling the device at a high speed or by simply probing the device. Devices that might utilize these PUFs are unmanned devices such as wireless sensor nodes deployed for infrastructure monitoring, environmental monitoring, or espionage missions. PUFs used in such attacks should have sufficient TTM security to prevent adversaries from modeling the PUF in a feasible amount of time. Table 9.1 summarizes the security classification we proposed.

Table 9.1 PUF-based viable security schemes and their suggested devices

CRP accessibility	Suggested devices	Restriction on area/power
Hidden	Low-power wearable computing or medical devices	High restriction
Limited	Passive RFID tags in ID cards or passports	Very high restriction
Full	Develop wireless sensor networks	Low/medium restriction

9.4   Implementation Results and Comparison

To identify the efficiency of the PUF designs in resisting modeling attacks, we implemented each design in 45 nm ASIC CMOS technology using the FreePDK45 digital library [144]. Synthesis results were generated by Cadence Encounter. The area requirement of each design is reported in the Gate Equivalent (GE) unit. The NCR and TTM values of each PUF were collected from [132] and [152]. Both of these values are then normalized according to the area requirement of the corresponding PUF design. This allowed us to produce their respective security efficiency:

1. NCR/GE: "Achieved NCR area efficiency." Indicates the PUF's ratio of achieved NCR over its area implementation cost.
2. TTM/GE: "Achieved TTM area efficiency." Indicates the PUF's ratio of achieved TTM over its area implementation cost.

It is noted that the security figures reported in [132] and [152] are not hard figures as they are based on experimental results. Nevertheless, these figures are accurate enough for comparing the security of each PUF particularly when the difference is significant. The result of the PUF implementations is inspected, and the security efficiency of each is evaluated.

9.4.1   Arbiter PUF

The NCR for the Arbiter PUF was obtained from [132]. The required implementation areas for all PUFs were obtained from our implementations. Table 9.2 shows the NCR Area efficiency of the Arbiter PUF for different challenge bit lengths. The area efficiency NCR/GE of the Arbiter PUF is constant as the NCR growth is linear.

The TTM values reported in [132] show polynomial growth for its efficiency. We compare the Arbiter PUF TTM efficiency with the rest of the implemented PUFs in Fig. 9.3.

Table 9.2 Arbiter PUF NCR efficiency values

Metric	Arbiter PUF		
Challenge length	64 bits	128 bits	256 bits
NCR	3300	6500	12,900
Area (GE)	351	698	1394
Efficiency (NCR/GE)	9.395	9.312	9.250

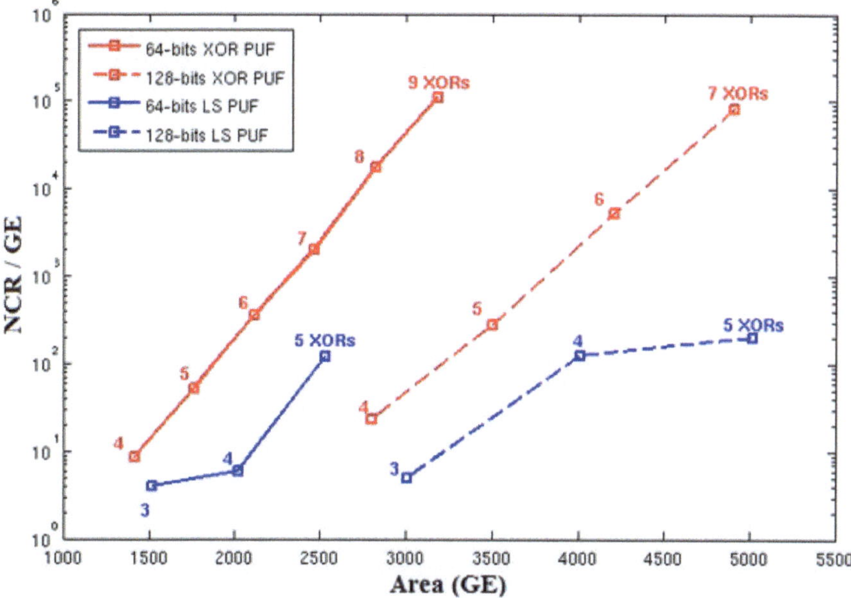

Fig. 9.1 NCR efficiency growth for the XOR PUR and the LS PUF

9.4.2 XOR Arbiter PUF

Figure 9.1 shows the NCR area efficiency of both 64-bit and 128-bit implementations of the XOR PUF. The NCR values were obtained from [152]. These values were also in accordance with the work performed in [132]. It is noted that the area efficiency increases as we increase the number of XOR-ed rows in the design. Figure 9.1 also shows that the 64-bit implementations of the XOR PUFs are more efficient than their 128-bit counterparts. For instance, a 64-bit, 8-input XOR PUF offers a much higher NCR value when compared to a 128-bit 4-input XOR PUF while consuming the same area on chip.

Figure 9.2 shows the TTM area efficiency growth of the implemented XOR PUFs. The TTM values were obtained from [152]. Similar to their NCR area efficiency growth, the TTM area efficiency also increases as we increase the number of XOR-ed rows. Again, the 64-bit implementations show a much higher efficiency when compared to their 128-bit counterpart.

9.4.3 LS Arbiter PUF

Figure 9.1 shows the NCR area efficiency of the 64-bit and 128-bit implementations of the LS PUF. The NCR values were obtained from [132]. Similar to the

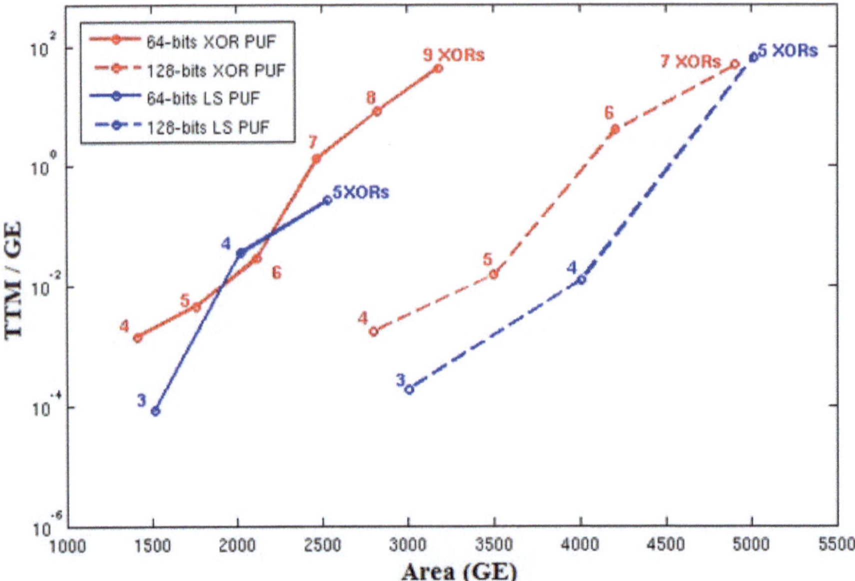

Fig. 9.2 TTM efficiency growth for the XOR PUR and the LS PUF

Table 9.3 3-loop FF PUF
NCR efficiency growth

Metric	Feed-forward PUF		
Challenge length	64 bits	128 bits	256 bits
NCR	19,353	37,567	73,997
Area (GE)	380	726	1423
Efficiency (NCR/GE)	50.92	51.74	52

XOR-Arbiter, the area efficiency increases as we increase the number of XOR-ed rows, with the 64-bit implementations being more efficient than the 128-bit implementations.

Figure 9.1 shows the TTM area efficiency growth of the LS PUF. The TTM values were also obtained from [132]. The efficiency increases with the number of XOR-ed rows while favoring the 64-bit implementations.

9.4.4   FF Arbiter PUF

Table 9.3 shows the NCR area efficiency of the FF PUF with different challenge bit lengths. NCR values were estimated based on the equations conjectured in [132]. The efficiency is almost constant as the equation suggests a linear growth with respect to the number of delay elements in the design.

We were not able to obtain values for the TTM area efficiency of the FF Arbiter PUF as there is a lack of such values in literature. However, Ruhrmair et al. in [132] suggested a polynomial growth of slightly higher order than that of the simple Arbiter PUF.

9.5 PUF Efficiency Comparison

Figure 9.3 shows a comparison between the implemented PUFs in terms of their NCR area efficiency. Both the arbiter PUF and the FF PUF are orders of magnitude behind the XOR and the LS PUF in terms of their security efficiency, which rather remains constant. This is due to the linear growth of their security when compared to the exponential growth of the XOR and LS PUFs. The FF Arbiter PUF does, however, offer a better NCR area efficiency when compared to the normal Arbiter PUF and has the best NCR area efficiency of all implemented PUFs for area restrictions below 1765 GE.

Comparing the NCR area efficiency of the XOR PUF with the LS PUF in Fig. 9.3, we find that the XOR PUF is more efficient. Although the LS PUF required the collection of more CRPs according to the model [132], when comparing it to the

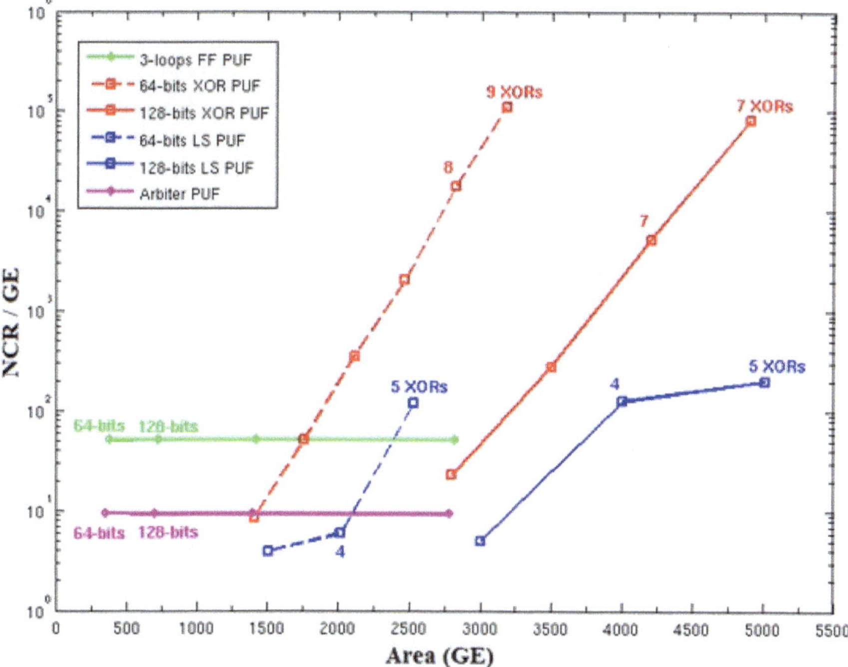

Fig. 9.3 NCR efficiency growth for all implemented circuits

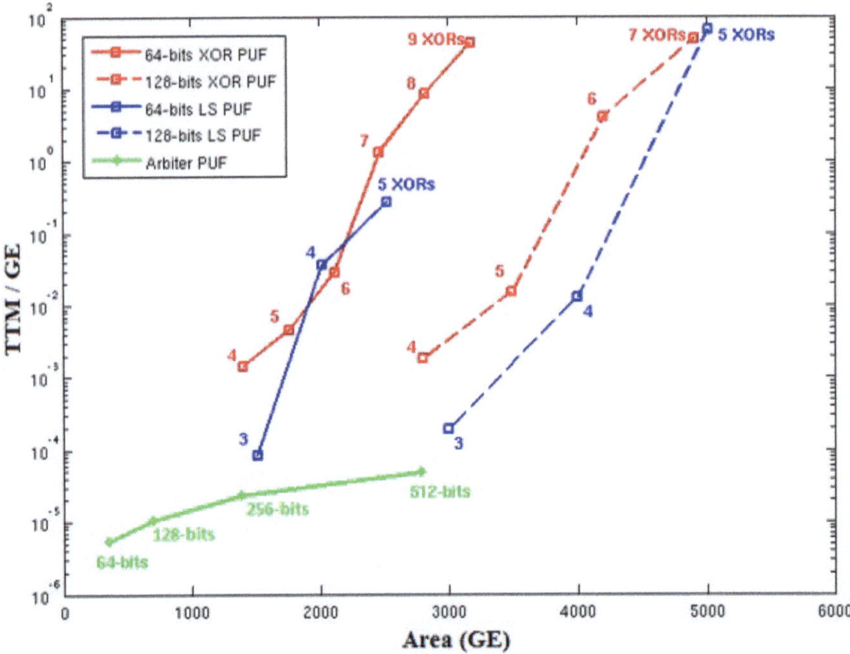

Fig. 9.4 TTM efficiency growth for all implemented circuits

XOR PUF in terms of its area efficiency, the XOR PUF is the clear winner. This is a rather important result as the LS PUF was introduced as an improvement over the XOR PUF. The area overhead imposed by its input network (calculated to be around 43% in our implementation) makes it much less efficient in regard to its NCR security. This additional area could be used to implement a larger XOR PUF capable of producing a higher NCR value.

Figure 9.4 shows a comparison between the TTM area efficiency of the XOR PUF and the LS PUF along with the Arbiter PUF. Surprisingly, the XOR PUF seems to have slightly better TTM efficiencies than the LS PUF. The difference, however, is rather marginal, considering that the values we obtained are based on the rather limited number of experiments performed in [132]. Still, such result was unexpected as the work in [132] reported a significant increase in the number of trials needed by the machine learning algorithm when modeling the LS PUF.

Figure 9.4 also shows how the Arbiter PUF severely lags behind the XOR PUF and LS PUF in terms of its TTM area efficiency.

9.6   PUF Design Implications and Use in Devices

The security efficiency values obtained through our work can serve as a guideline for PUF design. The efficiency values in Figs. 9.3 and 9.4 show that utilizing XOR output networks provides the most effective increase in security as opposed to utilizing XOR input networks in LS-PUF. The results also show that feed-forward loops are very efficient in providing increased security. The FF PUF, which utilized 3 FF loops, provided a significant increase in NCR security compared to the traditional arbiter with minimal area overhead. New PUF designs should incorporate the recognized effective elements and avoid the ones with minimal impact.

The security efficiency would also allow us to recognize the most suitable PUF to use for the different security schemes identified earlier. Table 9.4 summarizes the recommended choice of PUF to use in the different security schemes.

For the Hidden CRP PUF security schemes, the PUFs with the lowest area requirements are the most suitable, as the security values are irrelevant when no CRPs can be collected by an adversary. The FF PUF and the Arbiter PUF both have similar area implementation costs, albeit the FF PUF provides slightly better security. The FF PUF added security would have very little impact on the security of the system overall as the latter depends heavily on the cryptographic functions utilized to hide the CRPs. For this reason, we recommend the use of the Arbiter PUF that has the smallest area requirement.

For PUF security schemes with limited CRPs accessibility, the most suitable PUFs are the ones with the highest NCR area efficiencies. Such PUFs would produce the largest number of CRPs required to model the PUF. Inspecting our implementation results in Fig. 9.3, we find the XOR PUF to be the most efficient design when it comes to its NCR values. It is noted, though, that at very strict area constraints (<1765 GE), the FF PUF will be the most efficient in terms of its NCR security.

PUF security schemes that utilize accessible CRPs architecture will depend on achieving an extremely high TTM value that would practically prevent any kind of modeling attack. Both the LS PUF and the XOR PUF have similar TTM area efficiency. Nevertheless, in order to achieve the required TTM values, very large PUFs need to be implemented. In [132], an XOR PUF with 512-bit challenge length and 8-input XORs was found to be immune against modeling attacks. According to

Table 9.4 Recommended PUF circuit design for each security scheme

Metric	CRP Accessibility		
	Hidden	Limited	Full
PUF design criteria	Minimum area	Highest NCR Efficiency	Highest TTM efficiency
Recommended PUF	Arbiter PUF	XOR PUF (>1765 GE) FF PUF (<1765 GE)	LS-PUF/XOR-PUF

our implementation, such PUF would require 22,301 GE. A similar LS PUF would require 31,931 GE. Both implementations cannot be employed by RFID tags that can only use up to 5K logic gates for security. Such large PUFs are also well beyond what is considered suitable for "lightweight" IoT devices, and as such, it might be more reasonable to use standard security for such exposed devices.

Chapter 10
Advanced PUF Designs

Innovations in PUF designs continue to push the boundaries of security and performance in hardware authentication. This chapter explores cutting-edge advancements in PUF technology, presenting novel designs and architectures aimed at enhancing security, reliability, and versatility. By delving into the latest research and development efforts, we uncover emerging trends such as hybrid PUFs, reconfigurable PUFs, and machine learning-based PUFs, which offer new avenues for addressing existing limitations and adapting to evolving security threats. From mitigating vulnerabilities to improving scalability and efficiency, these advanced PUF designs hold promise for bolstering the resilience of hardware-based security solutions in an increasingly interconnected world.

10.1 A Digitized Memory-Based Implementation of the Delay-Based APUF

Weak PUFs are characterized by having a relatively small challenge-response. Such PUFs are mostly used as an alternative to secure key storage and overall have very high reliability. Their small challenge-response space, however, prevents them from being deployed as a cryptographic primitive in lightweight applications. Strong PUFs, on the other hand, have a large challenge-response space making them useful for authentication protocols. Strong PUFs are extremely sensitive to their input; while this allows them to establish a large and seemingly random challenge-response space, a crucial requirement for their use as lightweight cryptographic primitive, it also makes them highly susceptible to noise and bit-flips in their output. This has proved to be highly problematic as it prevented their security scaling and, ultimately, their adoption by commercial applications due to their vulnerability against machine learning-based attacks [7, 49, 131, 153].

© The Author(s), under exclusive license to Springer Nature Switzerland AG 2025

K. Khalil et al., *Lightweight Hardware Security and Physically Unclonable Functions*, https://doi.org/10.1007/978-3-031-76328-1_10

This chapter introduces a unique PUF design where we combine the benefits of Weak PUFs (i.e., high reliability and small area overhead) with the exponential challenge-response space of Strong PUFs while maintaining a minimal implementation area compared to traditional PUF designs. The introduced design is a unique and novel approach for implementing Strong PUFs where the strong PUF is implemented as an arithmetic module. At the same time, its entropy is stored in a reliable Weak PUF circuit. The system employs a simple hardware circuit that simulates the behavior of a strong PUF while using the Weak PUF circuit as its uniqueness source. The Strong PUF circuit is now a simple synchronous CMOS logic circuit, and its output is entirely reliable and resistant to noise effects. This approach ensures consistency of the output while also allowing for iterative binary operations to be performed on the PUF output without observing the avalanche effect of error accumulation. The result is a PUF primitive that has the reliability of Weak PUFs and the security and exponential challenge space of Strong PUF circuits.

10.2   Memory-Based Arbiter PUF Design Concept

Although unclonable through hardware, strong PUFs have been found to be vulnerable to modeling attacks. Various machine learning algorithms can be used to determine the separating hyperplane $\omega^T \Theta = 0$ that serves as the decision boundary surface for the response bit. Linear Regression (LR) was shown to be a very efficient algorithm in solving for ω. Once ω is determined using machine learning algorithms, it can be stored and used in Eq. 8.1 to produce an accurate value of Δ indicating whether the response of the PUF circuit is "1" or "0" depending on the sign (Δ). This method of generating a PUF response is highly reliable as it is fully digitized and not susceptible to noise, unlike the actual PUF circuit where the signal is raced along the paths. While a soft model can be used by most devices to generate the PUF output, it would be problematic to implement it in constrained devices, which are the target applications of PUFs. The soft response generation utilized by the machine learning attack uses floating-point matrix multiplication, which would be beyond the capacity of constrained devices. Moreover, using such a soft model would require storing the ω on the device, hence making it vulnerable to extraction through probing, which negates one of the key advantages of using a PUF circuit as opposed to a simple pseudo-cryptographic protocol with a stored key.

A digital implementation of the Arbiter PUF overcomes the need for traditional secret key storage and heavy computational circuitry. Utilizing a Weak PUF to store the value of ω retains the resistance against probing attacks. The heavy computation associated with Eq. 8.1 is avoided entirely by utilizing fixed-point calculation instead of floating point. While floating-point operations are required for performing a machine learning attack to generate a soft model, our experimentation has shown that a vector ω of fixed-point value would offer similar performance in terms of security, uniqueness, and randomness as, ultimately, the PUF output response is a fixed point $R \in (0, 1)$. Furthermore, the matrix multiplication is reduced to simple addition. This is possible since $\Theta \in (-1, 1)$. The end result is a compact PUF module with very similar security to that of the non-digital version while offering the important advantage of error-free outputs.

The introduced Memory-based Arbiter PUF has all the requirements for a Strong PUF: reliability, security against probing and modeling attacks, and a small implementation area. Its 100% reliability allows its security to scale indefinitely, securing it against modeling attacks. The Mem-APUF implementation retains the PUF's resistance to probing attacks by using a Weak PUF to store its source of entropy. Furthermore, its compact design is verified through implementation and shown to be suitable for constrained devices.

10.3   Mem-APUF Architecture

Figure 10.1 shows the architecture of the introduced APUF. An addressable Weak PUF is used to generate the ω vector that holds the differential delay values for the APUF. The Weak PUF of choice could be an SRAM PUF [60, 99] or LEDPUF [156], as both can achieve 100% reliability and have a small area overhead. The total number of bits needed to be stored by the Weak PUF would be 4 bits for each value of ω, which is equivalent to 4 bits for each of the APUF stages. The 4-bit

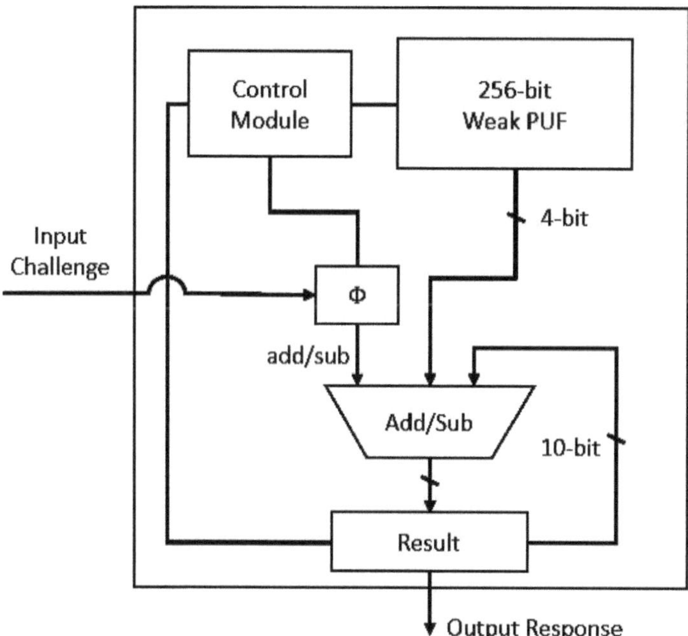

Fig. 10.1 Memory-based Arbiter PUF architecture

value was reached through experimenting with multiple values and observing the security, uniqueness, and randomness of the PUF. With 4 bits per ω value, a 64-bit input APUF (64 stages) would require $4 \times 64 = 256$ bits to store the full ω vector. This results in a minimal implementation overhead for the Weak PUF. The APUF arithmetic operation is implemented through a 10-bit, fixed-point, signed, serial adder/subtractor circuit that is fed with the values of ω one stage (4 bits) at a time along with the value of Θ corresponding to that stage. The value of Θ determines whether to perform an addition or subtraction. The vector Θ is generated serially from the input challenge C according to Eq. 8.2. The challenge transformation is straightforward and uses a single XOR gate along with a register. The final response R would simply be the sign bit of the final accumulated value.

The choice of serializing both the addition/subtraction and the Θ transformation significantly reduced the implementation area of our design while retaining the required level of accuracy and security. The Mem-PUF was implemented in 45 nm using Synopsys and Cadence tools (excluding the Weak PUF circuit) and showed an implementation area requirement of 280 GE, which makes it very suitable for constrained devices. For a 64-stage Arbiter PUF, observing 640 challenge-response pairs (CRPs) would allow an adversary to produce a soft model of the PUF with 95% accuracy and very little training time (<1 sec).

10.4 Mem-APUF's Weight Representation

In this section, we present the analysis and calibration of the Mem-APUF. A mathematical model of the Mem-PUF was used to derive its quality metrics: randomness, uniqueness, and its security in terms of resistance to machine learning attacks. The objective of the iterative design testing is to determine a suitable size for the differential delay values in ω. While increasing the number of bits for each value in ω would result in better quality metrics (uniqueness, randomness, and security), this suffers from diminishing returns and adds increasing storage requirement. The experiments showed that increasing the size of ω beyond 4 bits for each stage would result in negligible gains in security, uniqueness, and randomness.

10.4.1 Weight's Bit-Width vs. Randomness

Figure 10.2 shows the randomness achieved for the Mem-APUF for bit-width of ω values varying from 2 to 6 bits. The APUF's randomness is also shown in the same figure. The results show that there is no distinguishable gain beyond increasing ω bit requirements to more than 4 bits per value. The results also show that the overall randomness of the Mem-PUF is even superior to that of the traditional arbiter. This is due to the natural bias of the arbiter circuit that is present in the traditional PUF. Since Mem-PUF is entirely digitized and memory-based rather than delay-based, no

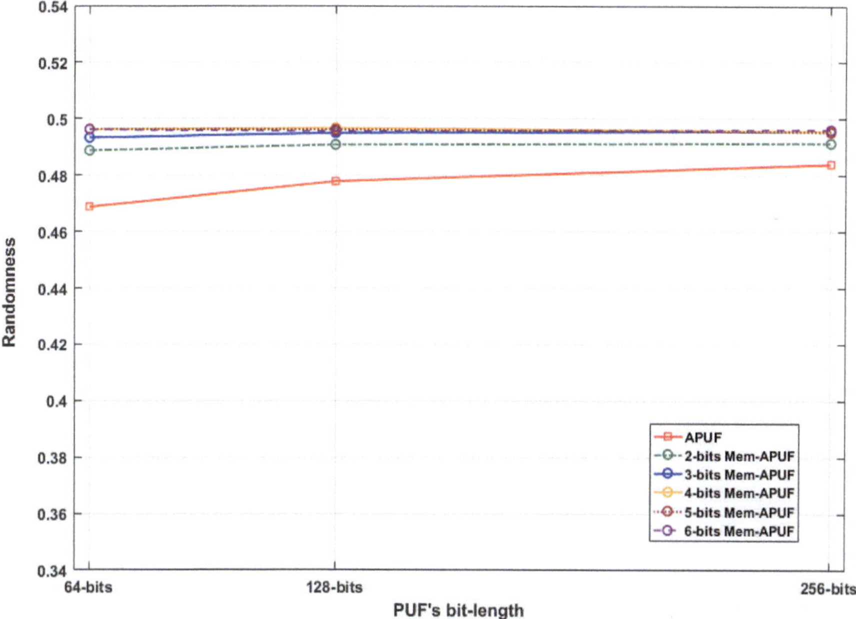

Fig. 10.2 The PUF's randomness at varying ω bit-width of a Mem-APUF compared to that of an APUF. Measured at varying PUF bit lengths of 64, 128, and 256 input sizes

such bias is present, and the overall randomness is superior to that of the traditional APUF.

10.4.2   Weight's Bit-Width vs. Uniqueness

Uniqueness is a measure of identifiability, where responses from different PUF instances should be distinguishable. Uniqueness of a PUF is measured by calculating the Inter-Chip Hamming Distance (Inter-Chip HD).

Figure 10.3 shows the uniqueness achieved by the Mem-APUF for bit-widths of ω varying from 2 to 6 bits. The gains in uniqueness beyond 4 bits seem to be negligible. However, the Mem-APUF achieved a slightly lower uniqueness value of 0.44 compared to 0.48 of the traditional APUF. This reduction in uniqueness proved to be inconsequential when we apply an XOR PUF [146] modification to the circuit, which is one of the popular ways to increase the randomness, security, and uniqueness of the Arbiter PUF.

Figure 10.4 shows that the difference in uniqueness values of a 4-bit XOR Mem-APUFs and traditional APUF becomes negligible beyond three XOR outputs. As the Mem-PUF is intended to have its security scaled up using XOR-ing or

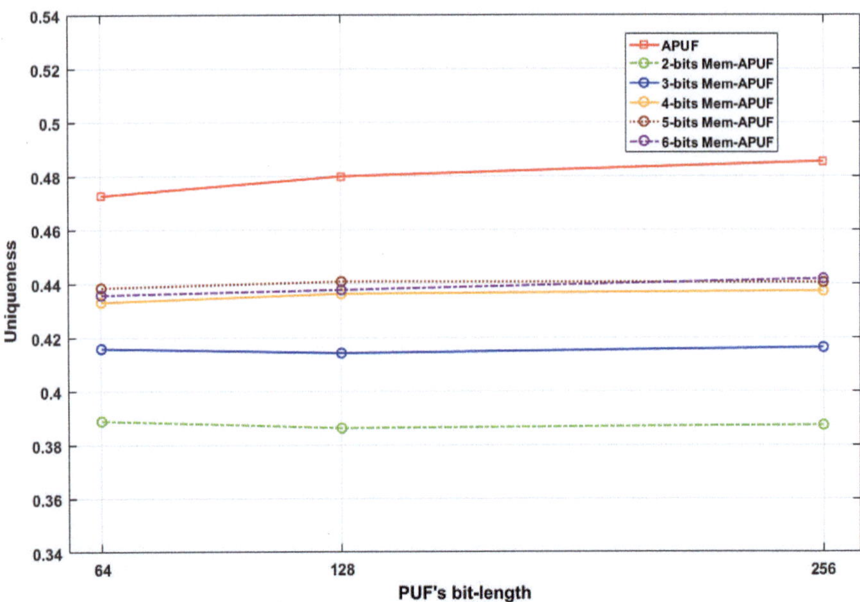

Fig. 10.3 The PUF's uniqueness at varying ω bit-width of a Mem-APUF compared to that of an APUF. Measured at varying PUF bit lengths of 64, 128, and 256 input sizes

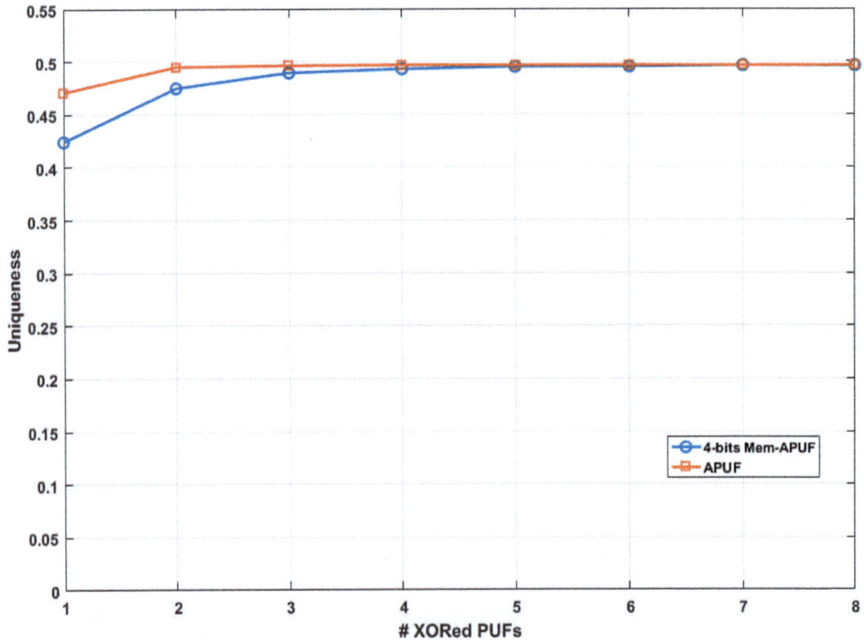

Fig. 10.4 Effect of XOR-ing multiple PUF instances on uniqueness when using a 4-bit Mem-APUF and an APUF

other techniques like Feed-Forward APUFs [92] or LS PUF[105], the achieved uniqueness by a 4-bit ω would be more than sufficient.

10.4.3 Weight's Bit-Width vs. Security

A PUF's resistance to modeling attacks is the cornerstone behind PUF security. This resistance is measured in NCR, the number of authentic Challenge-Response pairs that need to be observed by an attacker to achieve a certain prediction accuracy. We compare the prediction accuracy of a Linear Regression-based attack on the Mem-PUF with ω bit length values ranging from 2 to 5 bits as well as the traditional APUF, as shown in Fig. 10.5. The prediction accuracy achieved are all very similar apart from the 2-bit ω, which shows increased vulnerability to the attack. As such we can assume that any value greater than or equal to 3 bits for length of individual ω delay weights would ensure that the Mem-APUF would match the original APUF security.

The APUF security is often scaled up using XOR-ing, feed forwarding, or other types of circuits like the LS PUF; it is important to ensure that the introduced Mem-APUF can scale in a similar fashion. To determine a bit length for the weights in

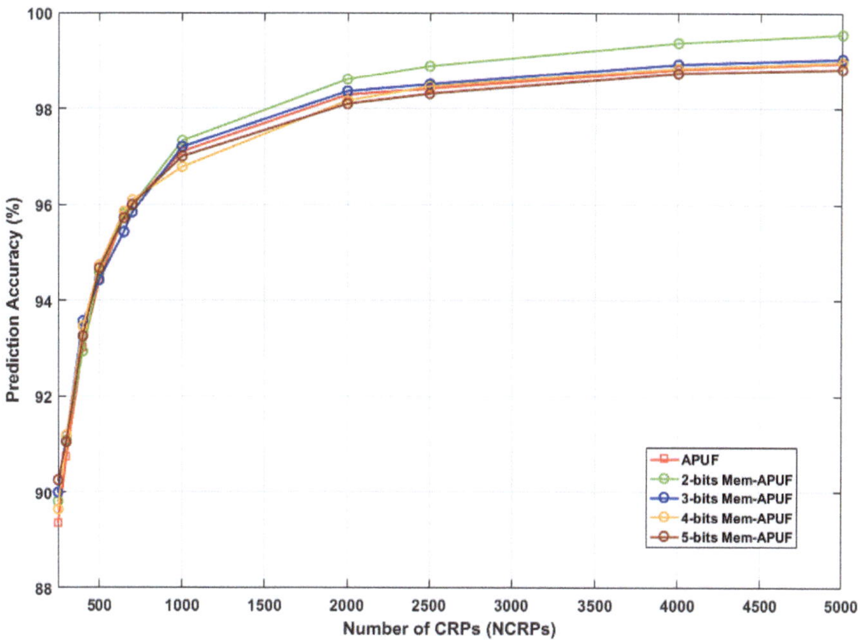

Fig. 10.5 Number of CRPs vs. prediction accuracy for varying Mem-APUF designs compared to an APUF

Table 10.1 Number of CRPs needed to reach prediction accuracies of 95% and 99% for different Mem-PUF bit-widths along with the APUF

3 XORs	3-bit Mem-APUF	4-bit Mem-APUF	5-bit Mem-APUF	APUF
(95%,99%)	(10k, 13.5k)	(20k, 45k)	(18k, 44k)	(22k, 45k)

ω that can ensure such growth, we implemented a scaled-up Mem-APUF circuit by XOR-ing the output from three instances of the module and performed modeling attacks utilizing neural networks. The results are compared with similar attacks we performed on a 3-XOR APUF and are shown in Table 10.1. From the table, we can see that by choosing a bit-width of 4 bits for the weights in ω, we can ensure comparable security scaling to that of a traditional APUF.

10.5   Mem-PUF's Summarized Features

While a value of 3 bits for the length of weights in ω would have sufficed for security, uniqueness testing showed that 4 bits would offer better uniqueness and is the threshold before diminishing returns offer little gain in bit length increase of ω. As such, a value of 4 bits for the length of the weights in ω seems to be the optimal one when considering all metrics: security, randomness, and uniqueness while taking into account the diminishing returns of further increase.

The Mem-APUF design elements are inspected and calibrated to ensure PUF's randomness and uniqueness metrics match those of the original APUF. The Mem-APUF security and its scaling are also verified using machine learning attacks and shown to achieve the desired scaling level. The introduced Mem-PUF would be the first to completely digitize a Strong PUF circuit like the APUF. The advantages of such digitization are clear: high reliability that would allow it to be employed in more complex authentication protocols and high scalability as combined instances would not suffer from error accumulation. This success in digitizing the most popular Strong PUFs would open the door for more digital, memory-based Strong PUFs as it shows that they can match the security and quality of delay-based Strong PUFs while maintaining a small implementation area overhead and offering a dramatic increase in reliability.

10.6   A Temporal Shadow PUF with Properties Isomorphic to the Delay-Based PUF

While ongoing research efforts continue to tackle the security of PUF-based authentication and key exchange protocols, an essential aspect of PUFs that has not been sufficiently investigated and addressed is their staticity. While devices utilizing traditional security primitives can easily be reconfigured with new secret keys, a PUF's defining characteristics are embedded in the hardware and cannot be changed. Exposing information about the PUF circuit can allow an adversary to model the PUF through various machine learning algorithms. Hence, once a PUF circuit has enough information exposed, its security might be permanently compromised. This shortcoming has been the subject of several research efforts that aim to offer non-static PUF constructs [38, 86].

This introduced work offers a novel technique based on information theory that allows PUFs to be used in non-static configurations. In the introduced technique, a Shadow PUF, a temporary PUF, is generated from the permanent PUF embedded in the device. The Shadow PUF is then used to perform a limited number of exchanges. The permanent PUF's behavior is never exposed as all authentications and exchanges are done using the Shadow PUF. The temporary "shadow" PUF is reconfigured periodically to avoid compromising information about the permanent PUF. The PUF's resilience to modeling attacks is analyzed to ensure the Shadow PUF and the permanent PUF's security. The Shadow PUF's reconfigurability can offer IoT devices reliable and lightweight authentication without the need to throttle device throughput or limit the number of authentications that can be performed.

10.7 Shadow PUF Architecture

The proposed Shadow PUF implements a reconfigurable, digitized version of the Arbiter PUF using fixed-point arithmetic. This approach is similar to the one used by the Mem-PUF for emulating the Arbiter PUF. However, while the Mem-PUF stores the values of its emulated delay weights in an SRAM-PUF, the Shadow PUF generates these weights dynamically and reconfigures them periodically.

Figure 10.6 shows the architecture of the Shadow PUF. The Shadow PUF weights are generated using a static Strong PUF circuit that would also be implemented in the device. The weight values of the Shadow PUF are never made public and are

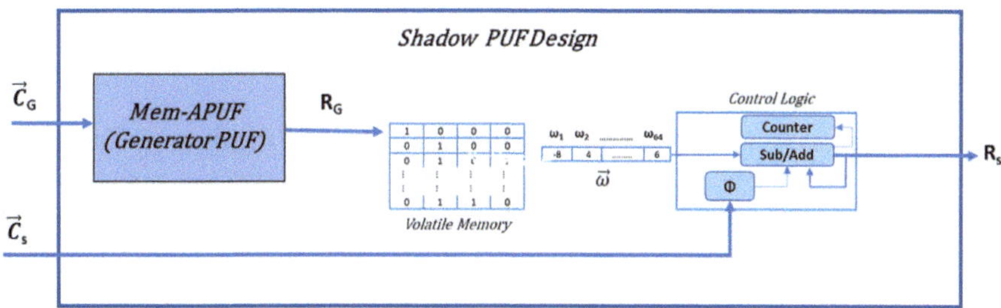

Fig. 10.6 Proposed shadow PUF architecture

only used for the generation of future responses. The weight values are periodically replaced with newly generated ones. A counter is used to track the number of evaluated Shadow-PUF CRPs and signal a request to generate new Shadow PUF weights. This limit on the CRPs generated using an individual set of weights prevents the shadow PUF from exposing enough information that could be used by machine learning-based attacks to compromise the system.

The choice of the utilized Strong PUF in our design, referred to as the Generator PUF, is not limited to any particular Strong PUF. However, any Generator PUF should have reliable responses as a single bit-flip in the generated weights can result in a corrupt and unpredictable Shadow-PUF module. In this chapter, the choice of the Generating PUF is the Mem-APUF. The discussed design assumes that the employed Mem-APUF's responses are 100% reliable. This reliability figure is easily attainable when an appropriate Weak PUF is utilized as a source of entropy for the Mem-APUF. Weak PUFs such as the NeoPUF [160] and the LEDPUF [156] have been shown to have ideal statistical properties and are capable of generating reliable responses that are resilient to aging effects without the use of additional error correction algorithms. S-PUFs utilizing fuzzy extractors are also a viable choice as they are also lightweight.

10.8 Shadow PUF Authentication Scenario

This section describes how the Shadow PUF can mutually authenticate the device and a trusted server. In this setup, the device would have a Mem-APUF circuit implemented, and the server would have access to a soft model of that circuit. The authentication steps are shown in Fig. 10.7. The server first sends a set of "generation" challenges C_g to the PUF device. These challenges will be fed to the static Mem-PUF, where their response would generate the weight of the Shadow PUF. Using the same challenges C_g sent, the server creates a Shadow PUF model using its stored Mem-APUF model. After the Shadow PUF's weight values are generated, the PUF device sends a set of randomly generated challenges to the server C_a and, at the same time, evaluates their responses Ra using the Shadow PUF. Finally, the server evaluates the responses R'_a of C_a and sends them back to be authenticated by the device if $R'_a = R_a$. This exchange would allow the device to authenticate the server.

For the server to authenticate the device, steps 3, 4, and 5 (as shown in the figure) are repeated. However, this time, the server would generate the authentication challenge and send it to the device to authenticate its response. A counter is used on the PUF device to keep track of the number of evaluated Shadow PUF responses to protect against ML attacks. When weights need replacing, the device would inform the other party, and a new set of challenges C_G must be exchanged.

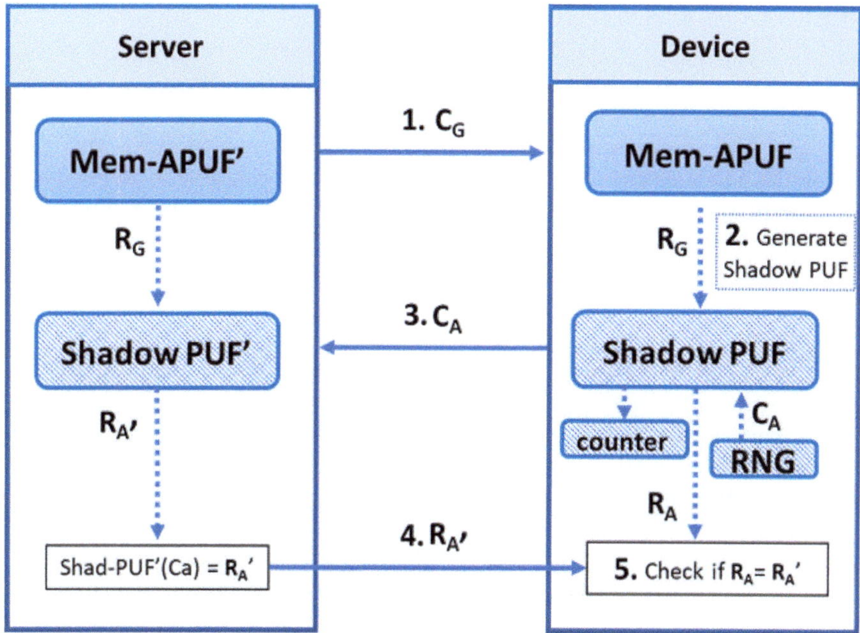

Fig. 10.7 An example of an authentication scenario utilizing a Shadow PUF

10.9   Shadow PUF Physical Characteristics

A defining feature of a PUF circuit is the unpredictability of its responses, which is reflected by its randomness and uniqueness attributes. As we are using the Mem-APUF to generate the weights of the Shadow PUF, we measure both PUFs randomness and uniqueness and show that there is no decline in the attributes of generated Shadow PUFs. In this analysis, both the Shadow PUFs and the Mem-APUF are 64-bit PUFs, and of their 64 weights, ω are represented by four binary bits. Using more than four bits will increase the design's area overhead with little benefit to its randomness and uniqueness. The PUF's characteristics are analyzed using a software model of the Mem-APUF and the Shadow PUF, which are based on 65 nm predictive CMOS technology [139]. The evaluation of each metric is done using 100 different PUF instances and using a pool of 10k randomly generated CRPs.

10.9.1   Randomness

A PUF should have high randomness to ensure there is no bias in its responses. Let P_1 be the bias of responses toward the value "1." Then, the randomness (H) of a

PUF device can be defined as follows [67, 105]:

$$H = \min(P_1, 100 - P_1), \quad \text{where } 0 \leq H \leq 50. \tag{10.1}$$

Figure 10.1 shows the randomness based on Eq. 10.1 for both PUFs. The Mem-APUF has a median and a mean randomness H of 0.488 and 0.4855; similarly, the Shadow PUF's randomness median and mean are 0.489 and 0.4875. Both PUFs show high randomness, close to an ideal value of 50%, as shown in Fig. 10.9.

10.9.2   Uniqueness

The uniqueness of a PUF is measured by calculating the Inter-chip Hamming Distance (Inter-chip HD). Let D be the normalized Inter-chip Hamming Distance (Inter-chip HD) between two PUF instances. Uniqueness (U) is defined as follows [67, 105]:

$$U = \min(D, 100 - D), \quad \text{where } 0 \leq U \leq 50. \tag{10.2}$$

In the case of the Shadow PUF, more comparisons between the Generator PUFs and the Shadow PUFs are necessary to ensure that there exists no correlation between the two. For this purpose, we introduce two new uniqueness metrics to evaluate the uniqueness of the Shadow PUF: inter-uniqueness and intra-uniqueness. Figure 10.8 provides an illustrative diagram of the three uniqueness metrics presented in this work.

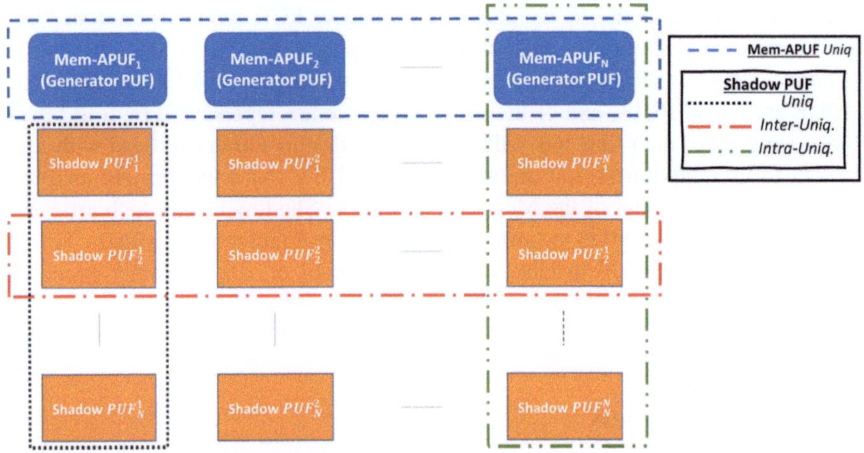

Fig. 10.8 A visual illustration of the utilized uniqueness metrics

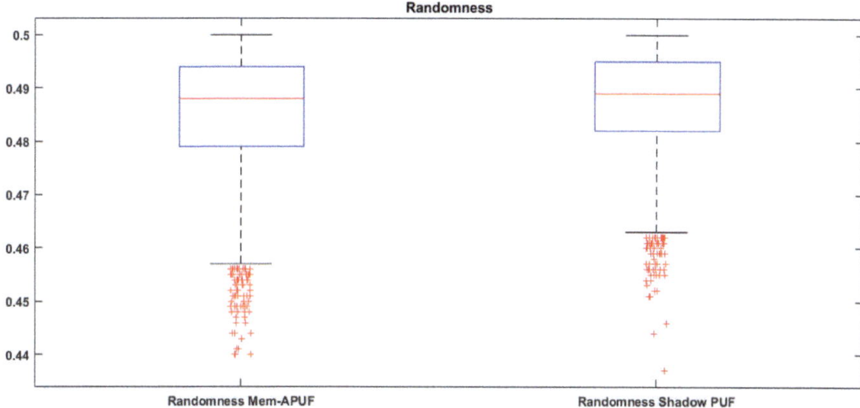

Fig. 10.9 Randomness of the Shadow PUF compared to that of the Mem-APUF

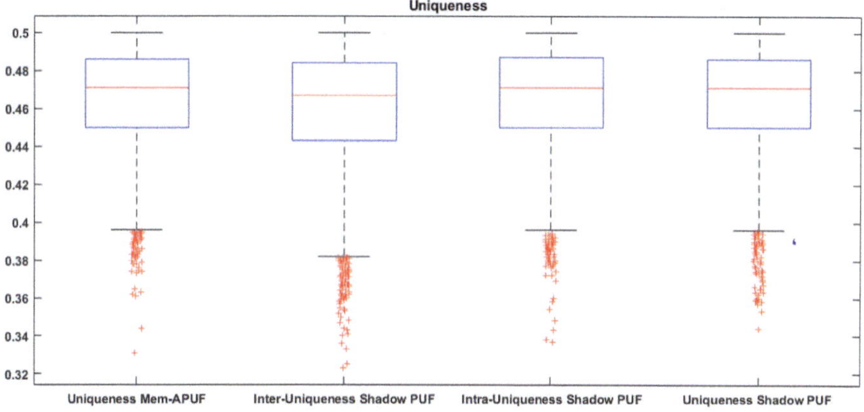

Fig. 10.10 Uniqueness of the Shadow PUF compared to that of the Mem-APUF

Inter-uniqueness: It answers the question of whether shadow PUFs are generated from the same challenge set but from distinct generators. Inter-uniqueness is measured by calculating the Inter-chip HD between Shadow PUFs that are generated from different Generator PUFs, but with the same set of challenges.

Intra-uniqueness: It answers whether different Shadow PUFs generated from the same Generator PUF are distinct from their parent. The Intra-uniqueness is measured by calculating the Inter-chip HD between the Generator PUF and all its generated Shadow PUFs.

The uniqueness of the Mem-APUF and the Shadow PUF is shown in Fig. 10.10. The lowest reported uniqueness is the Shadow PUF's Inter-uniqueness with median and mean U of 0.467 and 0.46, with a negligible difference to other measured uniqueness metrics. The results have shown that the Shadow PUF has high uniqueness, and it exhibited no loss compared to its generator PUF, the Mem-APUF.

Table 10.2 Implementation cost of the Shadow PUF design

Metric	Generator PUF	Shadow PUF		
		Memory[a]	Control logic	
Size	64 bits	256 cells	Using a 10-bit adder/subtractor	Total
Area (GE)	1583[b]	704	879	3166

[a] Memory is inferred by the synthesizer as latches
[b] If using a Mem-APUF architecture

10.10 Shadow PUF: ASIC Implementation

A Verilog behavioral-level design of the proposed Shadow PUF was synthesized in 45 nm [143] CMOS technology using Synopsys Design Vision. The implemented 64-bit Shadow PUF utilizes a 10-bit counter for TTL and a 10-bit adder/subtractor. The weight values of the Shadow PUFs are stored on a synchronous RAM (256 bits), which is inferred by the synthesizer as Latches and not D-Flip-Flops for a smaller area and power footprint. The Shadow PUF, running at 1000 MHz, has an area overhead of 1583 gate equivalent (GE) and a total power of 0.572 mW. The generator PUF and the Mem-APUF would have the same area and power estimate. The total area overhead that includes both the Mem-APUF and the Shadow PUF would be 3166 GE; Table 10.2 shows a breakdown of the area cost of the proposed design. In this chapter, we have utilized the Mem-APUF as the generator PUF; however, the generator PUF can possibly be replaced by other Strong PUF designs that might incur a smaller area footprint.

10.11 Shadow PUF Security Analysis

There have been many studies on the modeling resistance of the APUF [49, 131, 136, 153], where Deep Neural Networks (DNN) and Covariance Matrix Adaptation - Evolution Strategy (CMA-ES) have shown to be most successful. A single APUF or a Mem-APUF instance can be modeled with 95% accuracy using either a DNN or a CMA-ES approach, as they both share the same mathematical model. Consequently, the same would hold true for the Shadow PUF. To ensure its security against modeling attacks, the Shadow PUF must reconfigure its weights periodically to avoid exposing compromising information about its weight values. Hence, we must make certain that whatever information (i.e., CRP exchanges) is exposed by the Shadow PUF cannot be utilized by an adversary to model neither the temporary PUF nor the permanent PUF that is used for generating the reconfigurable weights. To analyze the security of the Mem-APUF and the Shadow PUF, we employ CMA-ES to model the PUF circuits as it is easier to infer the numerical weight values of its model after converging, especially if we are to utilize XOR-ing of multiple PUF instances. The results show that limiting the number of exchanges by the Shadow

PUF can successfully protect both the temporary PUF and the permanent PUFs from adversaries.

10.11.1   Modeling Attack on the Shadow PUF

We utilized the CMA-ES algorithm to predict the Shadow PUFs stored weights when observing its challenge-response exchanges. By predicting the Shadow PUF weights, we can predict the output of the temporary shadow PUF and expose the output response of the permanent Mem-PUF.

From Fig. 10.7, we recall that the weights of the Shadow PUF are nothing but the responses R_G of the permanent Mem-PUF when presented with generating challenge C_G. Hence, successfully predicting the Shadow PUF weights can expose the response of the permanent Mem-PUF associated with the device, allowing for further attacks to be performed on the permanent PUF itself. For this reason, we need to ensure that the shadow-PUF is reconfigured often to prevent any significant accuracy gains.

Figure 10.11 shows the Shadow PUF-trained model's accuracy when using various CRPs collected by the attacker. The figure also shows the prediction accuracy of the Shadow PUF's weights as the model converges to an accurate one. Looking at the results, we can observe that some weight values in the w vector can be predicted with higher accuracy than others when presented with a limited CPR set. This work is the first to report this significant observation, indicating that portions of the w vector can be compromised much earlier than the rest. These weights, in particular, are more impactful on the overall model's accuracy.

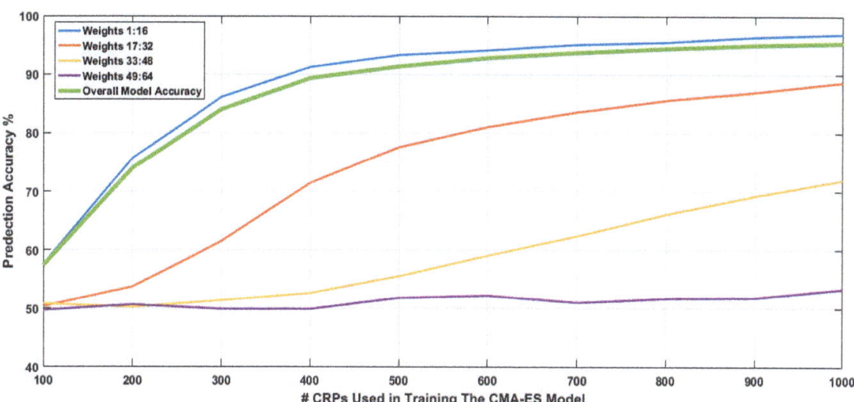

Fig. 10.11 The accuracy of the Shadow PUF's model and its weight values vs. the number of observed CRPs

10.11.2 Modeling Attack on the Generating PUF

When left without proper periodic reconfiguration, an average model accuracy of 95% for the Shadow PUF can be obtained by an adversary when observing 640 CRPs, as shown in Fig. 10.11. In our case study, the generator PUF would be the Mem-APUF, which has similar security to that of the Shadow PUF. Hence, to train an accurate model of Mem-APUF, an attacker must observe 640 CRPs for the Mem-PUF. However, as the Mem-PUF itself does not expose its responses, the attacker must resort to modeling the Shadow PUF and determining its weights, which are, in fact, the responses of the permanent Mem-PUF and can then be used in a second-stage attack aimed at modeling the permanent PUF. We assume that the attacker does not have access to the responses generated by the Mem-APUF, a reasonable assumption as the responses are never exposed and only temporarily used to generate a Shadow PUF on the device. Therefore, we aim to calculate the maximum number of responses that can be generated and exposed during authentication by the Shadow PUF without revealing any information that would allow the attacker to predict any of the Shadow PUF's weight bit values accurately.

10.11.3 Determining the Time to Live

The allowed number of evaluations of the Shadow PUF's Time to Live (TTL) is based on the minimum number of observed CRPs required to accurately predict at least one 1-bit value of the Shadow PUF's weight representation. As an example, Fig. 10.12 shows the prediction accuracy of all 64 4-bit weights of a Shadow PUF when training the model using (a) 100 CRPs and (b) 1000 CRPs. The first weight in a Shadow PUF is analogous to the last switch component's delay measurements in an APUF. This shows that an attacker can predict some of the bit values of the Shadow PUF's weights much earlier than others. From Fig. 10.12, we can also see that we must limit the CRPs exposed to 100 CRPs per temporary Shadow PUF model. After which, the Shadow PUF model must be reconfigured with new weights from the permanent Generator PUF. This ensures that the accuracy of the Shadow PUF model obtained through a CMA-ES attack cannot exceed 60%. It also ensures that none of the weights can be predicted with more than 60% accuracy. The TTL limit of 100 CRPs can be easily increased through XOR-ing multiple responses or incorporating feed-forward loops in our arithmetic operations. For example, XOR-ing two responses would increase the Shadow PUF TTL to 1250 CRPs.

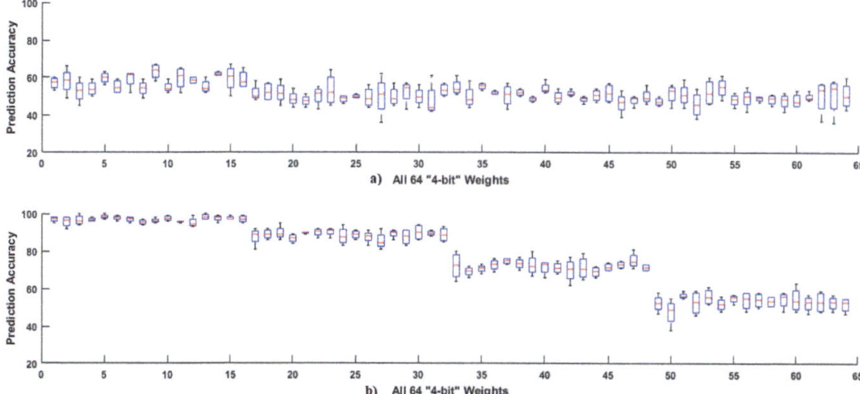

Fig. 10.12 The prediction accuracy of individual weights of the Shadow PUF at (**a**) 100 CRPs and (**b**) 1000 CRPs

10.12 Shadow PUF's Summarized Features

The Shadow PUF offers lightweight, secure communication for devices. Its weights are generated using a Memory-based Arbiter PUF associated with the device. By utilizing "weak" memory-based PUFs as a source for entropy, the Shadow PUF provides resistance to side-channel attacks used on the delay-based APUF. Analysis of the Shadow PUF showed that its randomness and uniqueness are near ideal and on par with traditional PUF circuits. Its weights are reconfigured periodically to ensure its resilience against machine learning attacks. A time-to-live metric for the weights has been produced based on experimental CMA-ES machine learning attacks. The TTL metric indicates how often the weights need to be refreshed to avoid exposing the PUF model. With its high reliability, the Shadow PUF can be easily scaled through XOR-ing and FF loops without fear of accumulating errors at the output. The Shadow PUF was implemented in ASIC to showcase its small area and low power requirements, making it suitable for lightweight devices.

Chapter 11
Lightweight PUF Protocols

As the demand for lightweight and efficient security solutions grows in the era of IoT and edge computing, the development of lightweight PUF protocols becomes paramount. This chapter explores the landscape of lightweight PUF protocols, focusing on their design principles, implementation strategies, and applications in resource-constrained environments. By leveraging the inherent uniqueness of PUFs while minimizing computational overhead and resource requirements, these protocols offer robust authentication and cryptographic functionalities suitable for small-scale devices with limited processing capabilities. Through a comprehensive examination of state-of-the-art lightweight PUF protocols, this chapter aims to provide insights into the evolving landscape of secure authentication mechanisms tailored for the emerging challenges of IoT and edge computing ecosystems.

11.1 A Lightweight Security Suite for Mutual Authentication and Secret Message Exchange

Although unclonable through hardware, strong PUFs have been found to be vulnerable to modeling attacks in [132, 152] and [48]. In such attacks, adversaries would collect the exchanged challenge-response pairs (CRPs) used in authentication sessions and apply machine learning algorithms to produce a software model of the PUF circuit. Such models would be capable of correctly predicting the response of any incoming PUF input challenge.

Consequently, PUF-based authentication protocols had to either employ very large PUF circuits to ensure resilience against modeling attacks or use cryptographic primitives to fully secure the challenge-response exchange, such as in [53, 134], and [154]. However, either solution would introduce significant implementation overhead, making it prohibitive for many low-power applications, and would ultimately defeat the purpose of utilizing PUFs. As an alternative to the computationally

© The Author(s), under exclusive license to Springer Nature Switzerland AG 2025
K. Khalil et al., *Lightweight Hardware Security and Physically Unclonable Functions*, https://doi.org/10.1007/978-3-031-76328-1_11

expensive standard cryptographic algorithms, PUF-based authentication protocols that use simple pseudo-cryptographic algorithms were suggested in [109, 167] and [51]. However, such protocols would offer less reliable security or limit the number of possible authentications to ensure resilience against modeling attacks.

To address this critical lack of a reliable, lightweight security solution, we have introduced a secure, lightweight PUF-based authentication protocol that can offer an unlimited number of mutual authentications. The Advanced PUF Protocol (APP) hides the PUF challenge-response correlations by conditionally transforming the challenges utilized. The protocol exchanges challenges that are generated by a true random number generator (TRNG) and then authenticates parties by checking for a required correlation between the exchanged challenges. The correlations under test are hidden using conditional transformation functions that transform the transmitted challenges dynamically prior to using them as inputs to the PUF circuit. The protocol's resilience against machine learning attacks is established by testing the performance of known attacks on the protocol, where we show its superior resilience nature.

We briefly list the main contributions we achieved under this research effort that aimed at introducing the first lightweight highly secure PUF protocol:

- We introduce a lightweight PUF-based mutual authentication and secret message exchange protocol based on a novel challenge-challenge exchange that can hide both the PUF responses and challenges. The protocol security and exceptional resilience against machine learning attacks is demonstrated by performing such attacks on a simulated setup.
- We introduce a lightweight PUF-based mutual authentication and secret message exchange protocol based on a novel challenge-challenge exchange that can hide both the PUF responses and challenges. The protocol security and exceptional resilience against machine learning attacks is demonstrated by performing such attacks on a simulated setup.
- We introduce novel challenge transformation functions that can transform the highly correlated challenges of Arbiter-based PUFs into multiple unique uncorrelated challenges while requiring a small implementation overhead.
- We present a security analysis of the protocol and compare its security with other recently introduced lightweight PUF-based authentication protocols to highlight the introduced protocol advantages.

11.2   Modeling Attacks and Arbiter PUF Variations

The Arbiter PUF was found to be vulnerable to modeling attacks in [132, 152] and [48]. In such attacks, an adversary would collect the exchanged challenge-response pairs (CRPs) used in authentication sessions and apply machine learning algorithms to produce a software model of the PUF that is capable of correctly predicting the response of new challenges.

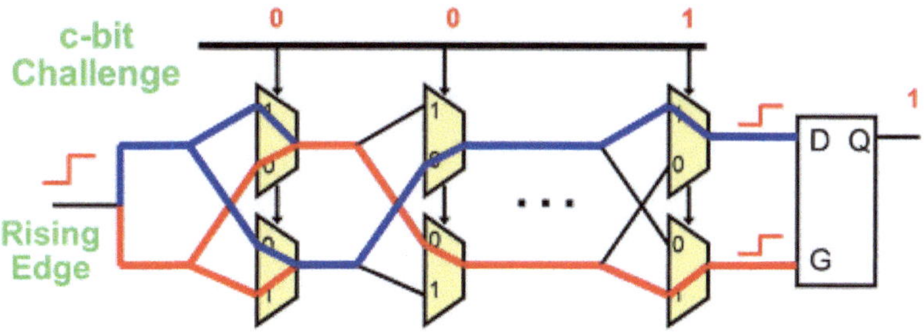

Fig. 11.1 Arbiter PUF design

The Arbiter PUF shown in Fig. 11.1 can be modeled as a set of delay elements. The delay difference at the arbiter Δ can be expressed as a function of the differential delay vector ω, and ϕ, the feature vector that is a function of the input challenge [132]:

$$\Delta = \omega^T \Phi. \tag{11.1}$$

Various machine learning algorithms can be used to determine the separating hyperplane $\omega^T \Phi = 0$ that serves as the decision boundary surface for the response bit. Linear Regression (LR) was shown to be a very efficient algorithm in solving for ω. For a 64-stage Arbiter PUF, observing 640 challenge-response pairs (CRPs) would allow an adversary to produce a soft model of the PUF with 95% accuracy and very little training time (<1 sec).

To address this vulnerability, several modifications of the Arbiter PUF have been suggested. The XOR-Arbiter PUF [147] combines several rows of the simple arbiter PUFs by XOR-ing their output into a single bit and introducing non-linearity in the PUF model. Other enhanced designs like the LS PUF [106] were suggested to increase the PUF resistance to modeling attacks as well. However, these introduced designs were also found to be vulnerable, albeit to a lesser extent, to modeling attacks in [132].

For highly nonlinear PUFs like FF-PUF [93], machine learning techniques utilizing Evolution Strategies (ES) were utilized to successfully produce a soft model of the PUF [132]. ES generates multiple offsprings of an initial model, which can be described by a delay vector w, using random mutations. The relative fitness of the offspring models is evaluated, and the fittest ones are used as new parents for the next generation. Evolution Strategy algorithms are particularly useful when the PUF challenge-response correlation is highly convoluted and linear regression would be infeasible to use. As long as the attacker can have a good measure of the relative fitness of the randomly generated models, the ES algorithm would eventually converge to the desired model.

It was suggested in [132] that modeling-resilient PUF designs would be possible to implement by drastically increasing the number of XOR-ed delay elements in some PUF designs. However, such PUFs have been shown in [70] to require a

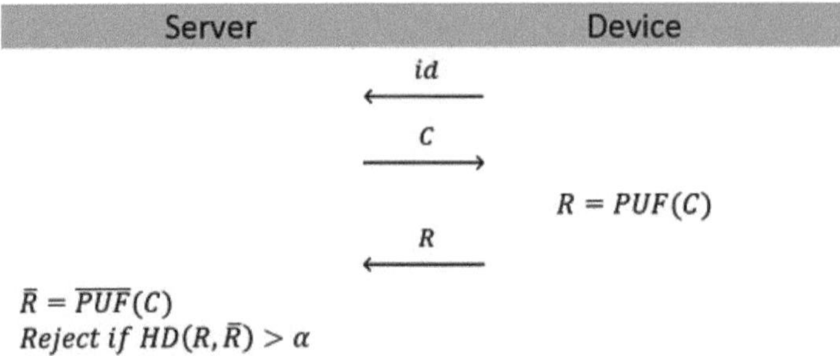

Fig. 11.2 A bare-bone challenge-response PUF-based authentication protocol

large implementation area ranging from 22k to 30k GE, making them infeasible to implement in constrained devices. With this vulnerability to modeling attacks, PUF-based authentication protocols had to employ other techniques to hide the challenge-response correlations of utilized PUF circuits.

11.2.1 PUFs in Authentication Protocols

Figure 11.2 shows a bare-bone PUF-based authentication protocol. A server with access to a soft model of the PUF would generate a set of challenges C and send it to the device. The device would use its PUF circuit to generate a set of responses R for the challenges in C and would send the responses back to the server. The server would compare the device's responses with the ones generated from the soft model. If the received and generated responses are within a certain error tolerance α, then the device would be deemed authentic.

The bare-bone authentication protocol described would be insecure as an adversary would be able to easily perform a modeling attack by collecting the exposed CRPs. Prior PUF-based protocols used varying methods to prevent such attacks. In [53, 134], and [154], cryptographic functions were used to hide the challenge-response exchange, providing reliable security albeit at the cost of using expensive hardware implementations that might not be suitable for constrained devices and hence defeat the purpose of PUF-based security.

To avoid the use of such expensive hardware, PUF-based authentication protocols that utilize simple pseudo-cryptographic algorithms were introduced in [109, 167] and [51]. These protocols, however, offer less reliable security or limit the number of possible authentications. In a later section, the security limitations of such protocols are examined in detail. The introduced protocol would eliminate all such disadvantages as it was designed with resistance to machine learning techniques in mind.

11.3 The Advanced PUF Protocol

11.3.1 Description of Used Notations

- Ci (ex: $C1, C2, ...$): indicates a single pseudo-challenge that can be used to generate an input challenge by applying a transformation function on it. We note that the challenge-challenge exchange utilizes a single pseudo-challenge $C1$ as the verifier's pseudo-challenge, and two pseudo-challenges $C2$ and $C3$ as the prover's response challenges.
- Ti (ex: $T1, T2, \ldots$): indicates a transformation function that transforms a single pseudo-challenge Tj into a transformed input challenge $Ti(Cj)$.
- $PUF(Ti(Cj))$: indicates the single bit response of the PUF circuit when presented with a transformed challenge $TiCj$. Only transformed challenges are fed to the PUF circuit.
- R_{Vi} (ex: $R_{V1}, R_{V2}, \ldots$): indicates a single PUF response bit, when presented with the transformed verifying challenge $Ti(C1)$.
- R_V (ex: $000, 001, \ldots$): indicates the concatenation of three response bits R_{V1}, R_{V2} and R_{V3}. In every exchange, three transformations $T1, T2,$ and $T3$, are applied on the pseudo-challenge $C1$ to produce R_V.

11.3.2 Challenge Exchange Overall Description

The introduced novel challenge-challenge exchange allows two parties to engage in an authentication without exposing any of the PUF circuit responses while ensuring resistance against machine learning attacks. Figure 11.3 shows how two parties, each with access to a PUF circuit or its soft model, can perform authentication while only exchanging random "pseudo" challenges. The exchange challenges are dubbed as "pseudo" challenges as they are never used as a direct challenge input to the PUF circuit but rather as a seed to new challenges that would be dynamically generated. A random pseudo-challenge $C1$ is presented by the verifier. The prover would receive the pseudo-challenge and would generate two random pseudo-challenges $C2$ and $C3$, such that:

$$PUF(Ti(C2)) \oplus PUF(Tj(C3)) = g, \tag{11.2}$$

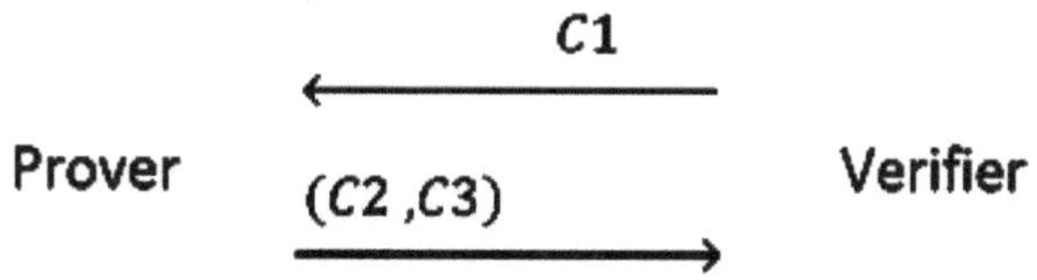

Fig. 11.3 The novel exchange can authenticate parties by exchanging randomly generated pseudo-challenges

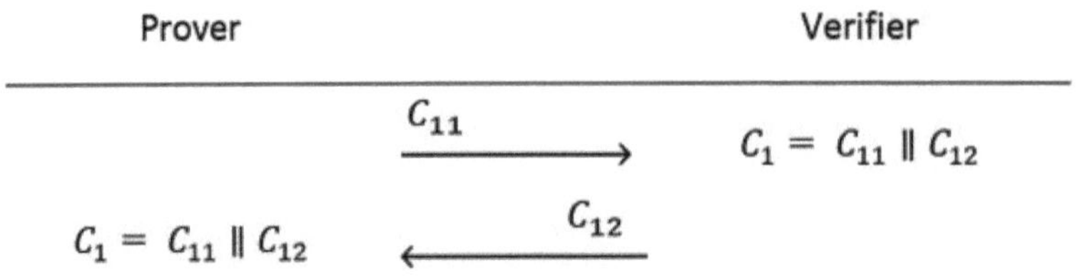

Prover   Verifier

$$\xrightarrow{\quad C_{11} \quad}$$
$$C_1 = C_{11} \parallel C_{12}$$

$$C_1 = C_{11} \parallel C_{12} \qquad \xleftarrow{\quad C_{12} \quad}$$

Fig. 11.4 Verifying challenge $C1$ is generated by both the prover and verifier using concatenation

where Ti and Tj are challenge transformation functions that transform a pseudo-challenge to a unique input challenge. The response values of the challenges generated from $C1$ would decide which of the transformations are used on the prover pseudo-challenges. The choice of the transformations applied is hidden from the adversary; hence, the actual PUF input is also hidden. The value of g is either 0 or 1 and is randomly selected by the prover at the start of each authentication round. The transformation steps are discussed in detail in the next subsection.

11.3.3   Combined Verifier-Prover Challenge Generation

Figure 11.4 shows how both the verifier and the prover generate a portion of the verifying challenge $C1$. For an n-bit challenge, each of the prover and verifier would generate (n/2)-bit challenges and exchange them. The verifying challenge would simply be the concatenation of the generated challenges.

This guarantees the freshness of each challenge exchange as it prevents adversaries from polling the device using the same set of challenges. However, such combined generation would reduce the effective challenge space of the PUF to $n/2$ bits as half of the bits can now be controlled by the prover. To maintain an effective challenge space of n-bit, the protocol would need to utilize a 2n-bit input PUF.

11.3.4   Conditional Challenge Transformation

Figure 11.5 shows the flowchart for the conditional transformation utilized. The verifier pseudo-challenge is transformed into three uncorrelated challenges, each producing a unique response. This would result in a 3-bit response for the versifier's pseudo-challenge. This response would be completely hidden from an attacker's perspective.

At the core of the protocol security is the conditional transformation of the challenges. This would hide the actual challenge input of the PUF circuit. For this reason, we refer to the challenges $C1$, $C2$, and $C3$ as "pseudo" challenges as they are never used as a direct input to the PUF circuit. The purpose of the transformation

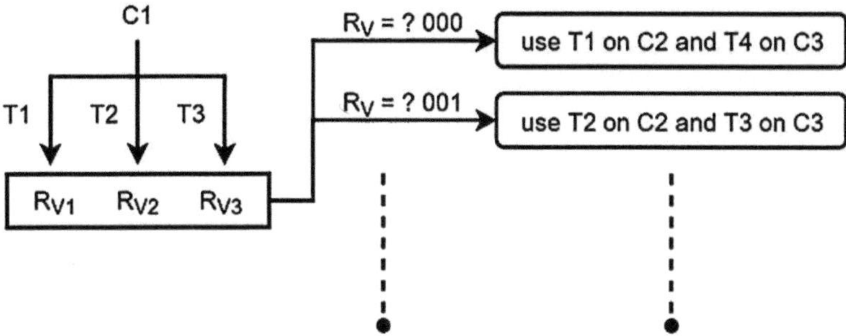

Fig. 11.5 Prover challenges $C2$ and $C3$ are transformed conditionally based on the verifying challenge $C1$ transformed responses

Table 11.1 The possible conditions tested according to the values of the responses generated from the verifying pseudo-challenge $C1$

R_v			Authentication condition
R_{v1}	R_{v2}	R_{v3}	PUF(TiC2) $\oplus$ PUF(TjC3) = g
0	0	0	PUF(T1(C2)) $\oplus$ PUF(T4(C3)) = g
0	0	1	PUF(T2(C2)) $\oplus$ PUF(T3(C3)) = g
0	1	0	PUF(T3(C2)) $\oplus$ PUF(T2(C3))= g
0	1	1	PUF(T4(C2)) $\oplus$ PUF(T1(C3)) = g
1	0	0	PUF(T1(C2)) $\oplus$ PUF(T3(C3)) = g
1	0	1	PUF(T2(C2)) $\oplus$ PUF(T4(C3)) = g
1	1	0	PUF(T3(C2)) $\oplus$ PUF(T1(C3)) = g
1	1	1	PUF(T4(C2)) $\oplus$ PUF(T2(C3)) = g

is to hide the PUF circuit input and greatly increase the non-linearity of the system. In the discussed version of the protocol, we utilize four uncorrelated transformation functions, $T1$, $T2$, $T3$, and $T4$, to transform the pseudo-challenges exchanged into new uncorrelated challenges.

Table 11.1 shows the eight authentication conditions that could possibly be under test by the verifier. The eight unique conditions are a result of applying a unique combination of transformations on $C2$ and $C3$. From an observer's perspective, any of these eight authentication conditions could have been under test. The choice of the transformations applied, and ultimately the authentication condition, is determined by the value of the responses of the verifier pseudo-challenge $C1$.

As g is a random value chosen by the prover, the verifier would have no access to its value. To test the authenticity of the prover, the verifier would simply check whether the pseudo-challenges presented by the prover in an authentication session would correspond to a consistent value of g. An adversary impersonating an authentic party would fail to produce a consistent value of g for all the m exchanges required as a proof of identity. The randomness of g adds little computational overhead to the protocol, while it significantly increases its security against machine learning-based attacks. This is discussed in more detail in a later subsection, where we demonstrate the protocol security against such attacks.

The design of four transformation functions that can guarantee to generate four uncorrelated challenges from a single pseudo-challenge can be a nontrivial task when utilizing some PUF circuits. In an ideal PUF, all challenge inputs would be completely uncorrelated. This would make the transformation function a very trivial challenge expansion function that simply uses the pseudo-challenge as a seed to generate three unique challenges.

However, in practice, some PUFs, like the Arbiter PUF and its variants, have a high correlation among their challenge inputs. In this chapter, we introduce a novel method for producing uncorrelated challenges when using the Arbiter PUF and its variants like the XOR PUF, LS PUF, or FF PUF. We note though that using such PUFs is not required by the protocol, we simply introduce this novel method to illustrate how such PUFs with a naturally high correlation between their input challenges can still be used in our protocol with very little overhead.

11.3.5　Authentication Protocol Steps

Figure 11.6 shows the authentication protocol utilizing the transformed challenges exchange. The verifier and prover would be parties in possession of the PUF circuit or an accurate soft model of the PUF that can be used to generate responses for any random challenge. As the verification procedure requires lightweight computation, both a PUF device and a standard device with a PUF soft model can authenticate the other party. This is one of the major advantages of the proposed protocol, as other protocols like Slender [109] or Challenge-obfuscation [51] require computationally heavy verification algorithms that are unsuitable to small devices and thus can only support one-way device authentication.

The protocol steps are listed below:

1. In step 1 of the protocol, an authentication session initialization message and the PUF ID are exchanged. The prover would generate a random value for g (0 or 1) that would be used for this session.
2. In step 2 of the protocol, the transformed challenge exchange is performed m times. As the verifier has no access to the value of g, they would simply save the value of $PUF(Ti(C1)) \oplus PUF(Tj(C3))$ in g'.
3. In step 3 of the protocol, the verifier would check whether the value of g' is the same in all the m exchanges performed. If not, it would reject to authenticate the device. The protocol can be modified to allow for error tolerance by allowing for mismatches in a small portion of the generated g' values. Such tolerance rate should be calibrated in accordance with the utilized PUF's circuit error rate.

An adversary would be unable to produce a fixed value of g for all the m exchanges in the authentication session. The probability of randomly generating such pseudo-challenges would be equal to $2 \times 2^{-m} = 2^{-(m-1)}$, which is equivalent to the probability of randomly producing a g value of 0 for all m exchanges or a g value of 1 for all m exchanges. This guarantees an exponential level of security

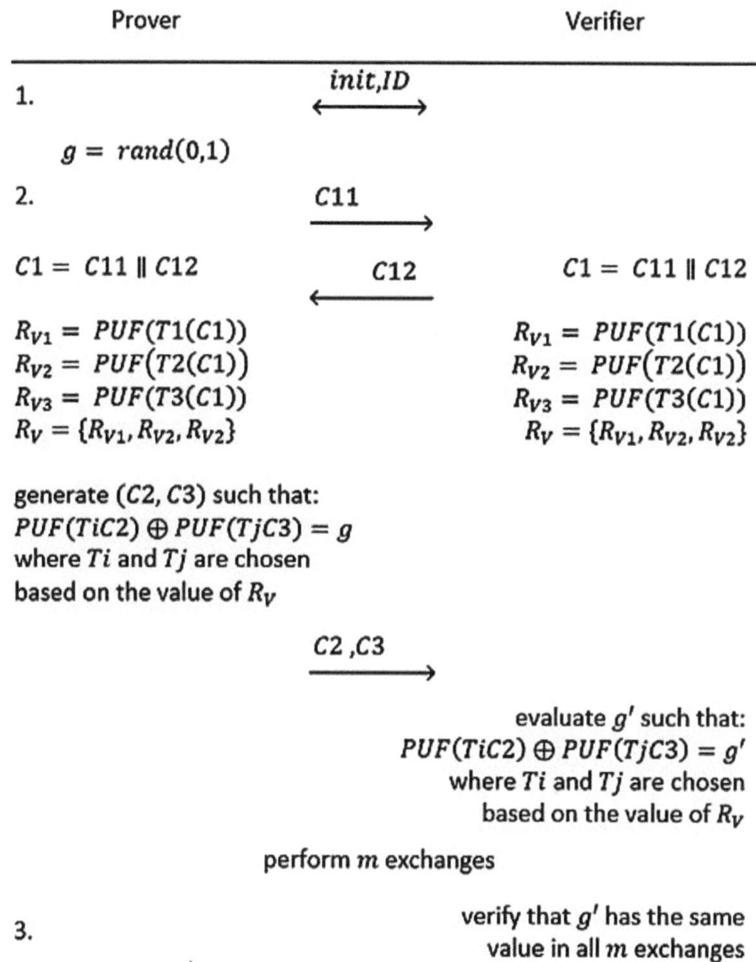

Fig. 11.6 Detailed steps of the advanced PUF protocol

against random guess attacks. We further investigate the security of the protocol against modeling attacks and challenge collection attacks in Sect. 11.6 (Fig. 11.7).

11.4 Secret Message Exchange

The enhanced challenge-challenge exchange can support secret message exchange. By having two possible authentic response values for a fixed value of g, we can encode each response with a different data bit. An adversary could only guess the actual data bit as the response of a PUF challenge is never exposed. Table 11.2 shows

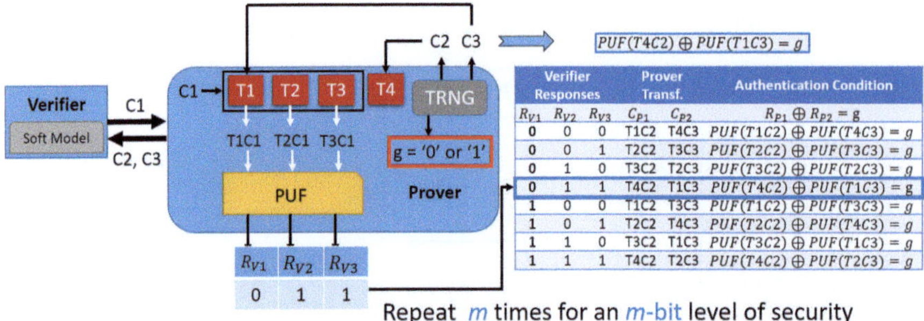

Fig. 11.7 An illustration of the protocol challenge-challenge-based authentication

Table 11.2 Results of a secret data exchange using the introduced protocol $C1$

$C_2,C3 \longrightarrow$			Authenticity (for g =0)	Data bit d_i
PUF $(T_i(C_2))$	PUF$(T_i(C_3))$	$\oplus$		
0	0	0	Authentic	0
0	0	0	Rejected	N/A
0	0	0	Rejected	N/A
0	0	0	Authentic	0

how the two possible authentic combinations can each be encoded with a different data bit in a session where $g = 0$.

11.5   Design of Uncorrelated Challenge Transformation Functions

In an ideal PUF, all the challenge inputs would be completely uncorrelated. This would make a challenge transformation function very simple. Each transformation function would add a unique bit string to the pseudo-challenge, resulting in a unique challenge associated with the transformation and the pseudo-challenge used. However, in practice, some PUFs, like the Arbiter PUF and its variants, have a high correlation among their challenge inputs, and as such, unique challenges are not necessarily uncorrelated [107]. This was one of the issues addressed in the design of the LS PUF [106].

In this chapter, we introduce a novel method for producing uncorrelated challenges when using the Arbiter PUF and its variants like the XOR PUF, LS PUF, FF PUF, or DA-PUF [69]. We note, though, that using such PUFs is not required by the protocol; we simply introduce this novel method to illustrate how these PUFs with a naturally high correlation between their input challenges can still be used in our protocol with very little overhead. In this section, we first examine the impact of induced challenge bit-flips on the response of an Arbiter PUF. We then explain

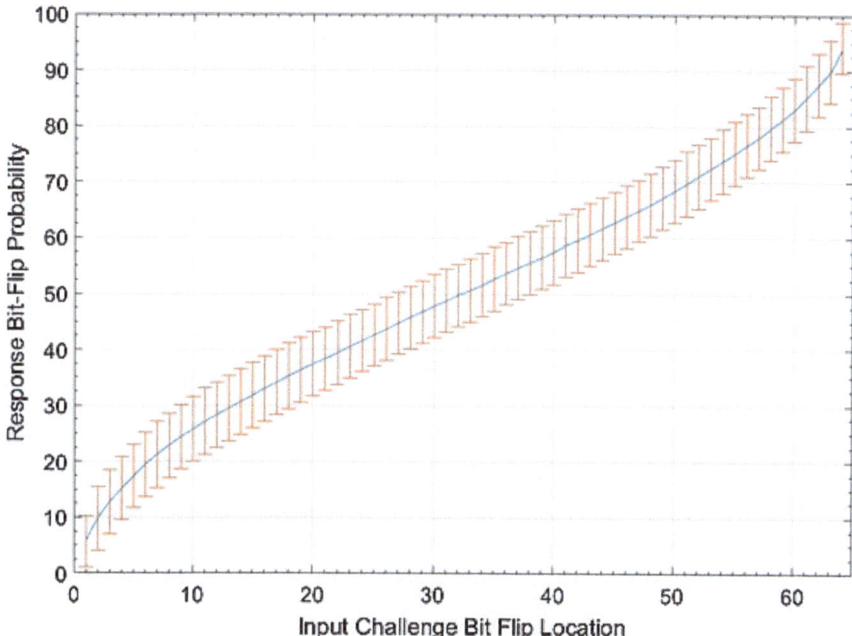

Fig. 11.8 The probability of causing a bit-flip in the response when an input challenge bit is flipped for a 64-bit arbiter PUF

how multiple input challenge bit-flips can be used to design uncorrelated challenge transformation functions.

11.5.1 Input Challenge Bit-Flips Response Correlation

The authors in [107] examined the effect of introducing a single bit-flip in the PUF input challenge on the output response of an arbiter PUF. We verified and reproduced their work using an arbiter PUF mathematical model. We used the same delay arbiter PUF model utilized in [107], which is based on the delay variations presented in [140]. The results are shown in Fig. 11.8, where we can see that input bit-flips introduced near the middle of the PUF circuit have a 50% chance of producing a bit-flip at the PUF output.

Prior to this work, the effect of multiple induced challenge bit-flips on Arbiter PUFs has not been properly examined. We have observed that introducing multiple bit-flips would have a drastically different effect than that of a single bit-flip. For instance, when using the same mathematical model utilized to produce Fig. 11.8, introducing two bit-flips at locations 31 and 33, respectively, would result in a 10% probability of obtaining a response bit-flip. This very low probability is contrary to

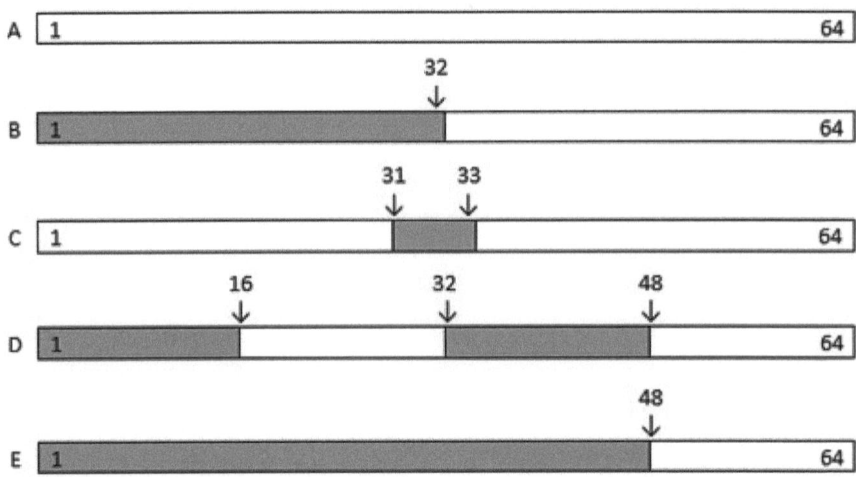

Fig. 11.9 The tested bit-flip instances and their effect on the delay elements path connections. Shaded areas represent delay elements that have swapped paths

the intuition of Fig. 11.8, where both bit-flips showed a probability very close to 50%.

To understand the PUF output correlations when multiple bit-flips are introduced, we examine a case study where five instances of input bit-flips are introduced, into a 64-bit PUF circuit. Figure 11.9 shows the effect and location of the introduced bit-flips instances A, B, C, D, and E on a 64-bit Arbiter PUF circuit.

The shaded areas in each instance show the portion of the PUF where the delay elements have swapped paths in the circuit (upper vs. lower) due to the introduced challenge bit-flips. Delay elements are either connected to the same path prior to the induced bit-flips (non-shaded) or swapped their path as a result of the induced bit-flips (shaded). We conjecture that the probability P_S of two instances of induced bit-flips to share the same PUF's response value can be predicted by the ratio of matching delay element connections over the total number of delay elements. When the shaded delay elements ratio is around 50%, the probability of producing a matching response should be around 50%.

To verify our conjectured estimation method, we used the mathematical model of the Arbiter PUF to evaluate the output of random challenges when the bit-flips instances shown in Fig. 11.9 are utilized. A total of 200 PUF instances were produced, and 5000 challenges tested in each instance. Table 11.3 shows the experimental results of the bit-flips instances along with the ratio of shared delay elements.

The results show that, indeed, the probability of introducing a response bit-flip is quasi-proportional to the ratio of shared delay elements. It can be seen that a 50% mismatch ratio would always yield a 50% probability of sharing the same response. For ratio values different than 50%, the relationship seems to be quasi-

Table 11.3 The tested instances matching connection ratio and the associated probability of producing the same output response bit

| | Correlations with instance A | | | |
	AB	AC	AD	AE
Mismatch ration	0.50	0.95	0.5	0.25
Probability P_s	0.505	0.885	0.501	0.338

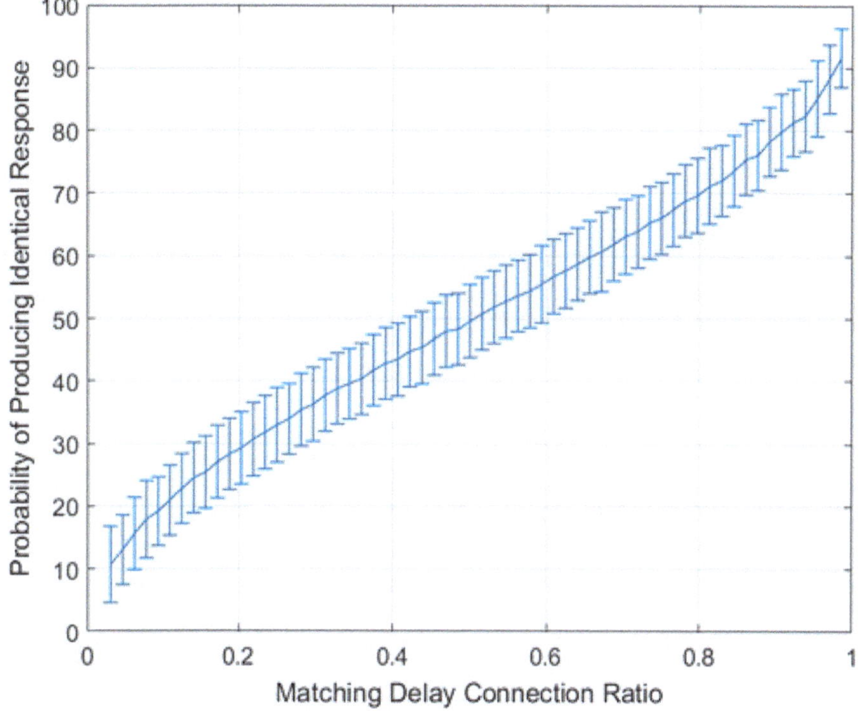

Fig. 11.10 Probability of response bit-flip correlation with the ratio of matching delay elements

proportional rather than proportional. The full relationship between the ratio of matching connection and output response is shown in Fig. 11.10. The data was produced by introducing multiple bit-flips at random locations and examining the response correlation with various challenges.

From the figure, we can see that an uncorrelated response can always be produced by introducing challenge bit-flips that result in a 0.5 ratio of mismatched delay elements connections. This insight is used in the next subsection to design four uncorrelated transformation functions.

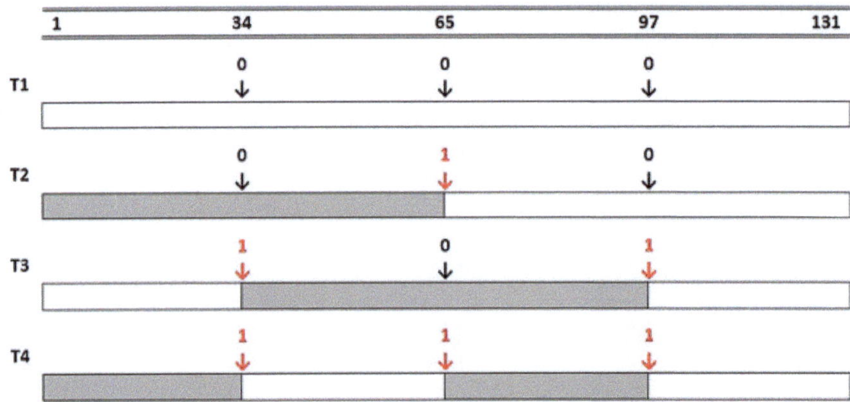

Fig. 11.11 The transformation functions T1–T4 are designed by inserting additional stages in the Arbiter PUF

Table 11.4 The transformation functions cross-correlation probabilities produced statistically from the PUF model

Transformation	T_2	T_3	T_4
T_1	$\mu = 0.499\ \sigma = 0.042$	$\mu = 0.487\ \sigma = 0.046$	$\mu = 0.498\ \sigma = 0.039$
T_2		$\mu = 0.497\ \sigma = 0.038$	$\mu = 0.486\ \sigma = 0.044$
T_3			$\mu = 0.498\ \sigma = 0.041$

11.5.2   Design of Uncorrelated Transformation Functions

Figure 11.11 illustrates how the padding bits chosen for the transformation functions would result in four unique circuit connections that each have a matching delay element connection (shaded/unshaded) ratio of 0.5 when compared to any other transformation instance. This ensures that there would be no correlation between any of the transformed challenges. For a protocol utilizing 128-bit pseudo-challenges and a 131-bit arbiter PUF circuit, these locations would be 34, 65, and 97.

The correlation among the transformation functions was verified using the Arbiter PUF mathematical model. The result of the verification is shown in Table 11.4, where 200 Arbiter PUF instances were fed with 5000 random pseudo-challenges. We can see that the functions are uncorrelated as the probability of producing the same response would be very close to 0.5.

11.6 Protocol Security Analysis

We analyze the security of the Advanced PUF-based Protocol for Lightweight Mutual Authentication against various attacks. We assume the protocol employs an Arbiter PUF with an n-bit input challenge and utilizes m exchanges in each authentication session. For simplicity, we ignore the error margin α as it can be addressed by increasing the number of exchanges m, as we show in a later section that examines the effect of the PUF error rate on the protocol security.

11.6.1 Random Guessing

In a random guess attack, the attacker simply responds with randomly generated pseudo-challenges. The probability of success for a random guess attack would be $2^{-(m-1)}$. A relatively small value of m would be sufficient for most applications as the attack does not expose any information or leak the model of the PUF.

11.6.2 Modeling Attacks

As the protocol hides both the challenges utilized and the PUF circuit response, it would protect the PUF circuit from modeling attacks that require direct access to the challenge-response pairs of the circuit. For each exchange, an attacker would have to guess which of the eight possible transformations occurred. The number of possible guesses required would grow exponentially when trying to perform such attacks. Machine learning attacks that utilize linear regression would fall into this category as they require a direct correlation between the circuit input and the output.

However, machine learning attacks utilizing the Evolution Strategies (ES) algorithm can be performed without such direct access to the challenge-response pairs. ES would generate random models and test their fitness according to the collected data. As long as the attacker can perform a "fitness" evaluation for a random model, such attacks can still be performed. In [10], ES was used to successfully generate an accurate model of the PUF circuits utilized by Slender. To evaluate the security of the proposed protocol against ES, we performed a Covariance Matrix Adaptation (CMA-ES) using a local implementation of the algorithm presented in [61].

We also performed the CMA-ES attack on Slender [109] to evaluate the advantage and the relative security gains of our introduced protocol. Slender protocol hides the response of the PUF circuit in "plain sight" by concatenating the PUF responses with a random bit stream generated by a TRNG and then logically rotating the string by a random number to hide the index of the authentic response substring. The server receiving the full string would then perform a substring matching algorithm to determine whether the authentic substring is part of the device response. The introduced protocol showed a solid resilience to CMA-ES attacks even when utilizing the notoriously vulnerable Arbiter PUF as the choice of the PUF circuit. We note how the CMA-ES was quickly able to compromise the PUF circuit model utilized by Slender.

We utilized a local implementation of the CMA-ES algorithm presented in [61]. The fitness function used in the attack against Slender measured the alignment score of the generated string and the collected string. This was done by rotating the collected strings and measuring the hamming distance from the tested model's generated string after each rotation. The normalized average of the minimum hamming distance among the strings under test was used as a final fitness value. Such fitness evaluation would give a very good measure of the relative model's accuracy in the pool, and this allows the ES algorithm to converge confidently toward more accurate models.

In the attack against our introduced protocol, the fitness function would evaluate the performance of the random PUF models by measuring the consistency of the g value in each round. A bias toward either value of $g = 0$ or $g = 1$ would correspond to a better fitness score regardless of the true value of g. Here, we see how the randomness of the g value in each session plays a critical role in confusing the ES algorithm, particularly when the model is around 50% accuracy. While an authentic round might have a g value of 0, a bias toward a g value of 1 would also give the model a better fitness score, although the tested model would have less accuracy.

In the ES attacks on the simulated setups, APP utilizes a 131-bit Arbiter PUF circuit and a number of $m = 100$ exchanges per authentication. Slender utilized a 131-bit Arbiter PUF and a full string length of $L = 2500$ and a substring of $L_S = 1250$ that is a similar setup to the one used in [109]. The simple Arbiter PUF is notorious for its low security and vulnerability, yet it was selected for testing purposes as its vulnerability would allow for a more noticeable impact on the protocol security rather than the PUF circuit security. For the attack on Slender, we utilized a pool of 1.25 million challenges, which formed 1000 observed authentications.

The ES algorithm randomly selected 100 observations, equivalent to 125,000 challenges in each iteration to evaluate the fitness function. For the attack on the introduced Advanced PUF Protocol (APP), we utilized 1 million observed exchanges, equivalent to the observation of 3 million pseudo-challenges or 10,000 authentication sessions when using $m = 100$ exchanges per authentication session. The ES would select 2000 random authentication sessions containing 200,000 exchanges for each iteration to evaluate the fitness function. Table 11.5 summarizes these values.

Figure 11.12 shows the achieved model accuracy progression of several runs over 120 generations of ES attacks on Slender (blue) and the proposed APP (red). We can see clearly how most ES attacks quickly converge on an accurate PUF model when utilizing Slender, while all the attacks on APP (red) failed to make any noticeable accuracy gains on APP even for runs initialized with relatively high accuracy. The normalized fitness function values and their progression are shown in Fig. 11.13.

Table 11.5 Number of authentications observed and challenges collected for each ES attack performed on Slender and APP

	Total pool size		Iteration sample size	
	Challenge collected	Authentication observed	Challenged collected	Authentications observed
Slender	1.25 million	1000	125,000	100
APP	3 million	10,000	200,000	2000

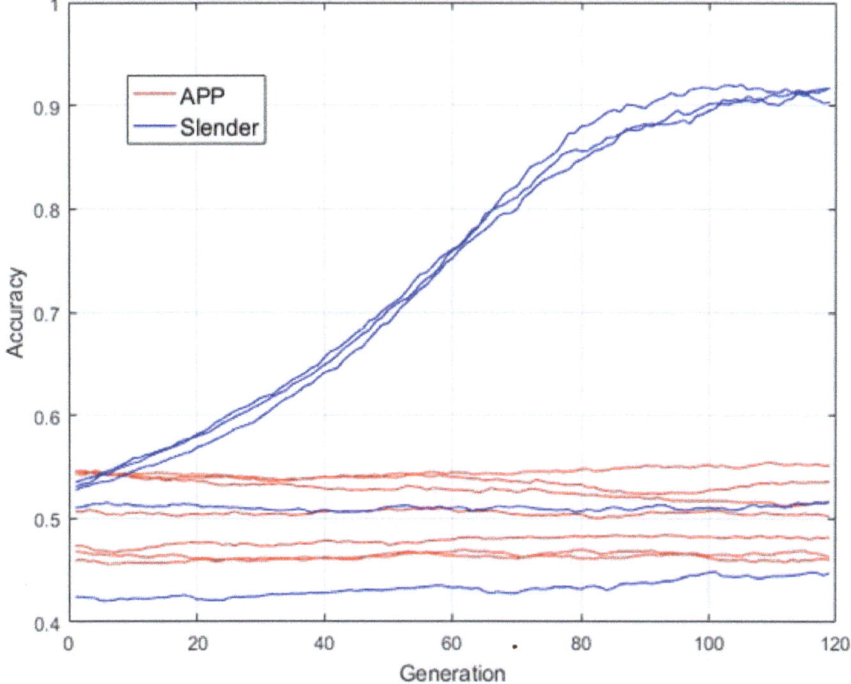

Fig. 11.12 Accuracy levels achieved by the CMA-ES attacks on Slender (blue) and on the introduced APP (red)

The figure also shows how the CMA-ES algorithm was very effective in minimizing the fitness value F on Slender (blue), while the fitness value of APP (red) did not show a noticeable decrease.

Figure 11.14 shows the accuracy levels attained by prolonged ES attacks on APP. We utilized custom-built initial accuracy conditions to inspect the impact of such initial conditions on the attack. The prolonged runs show that there would be no accuracy gains as long as the initial accuracy is below 60%. The ES attack with the highest initial accuracy showed a noticeable increase in accuracy after 400 generations, and we conjecture that with a longer run it could have converged to an accurate model.

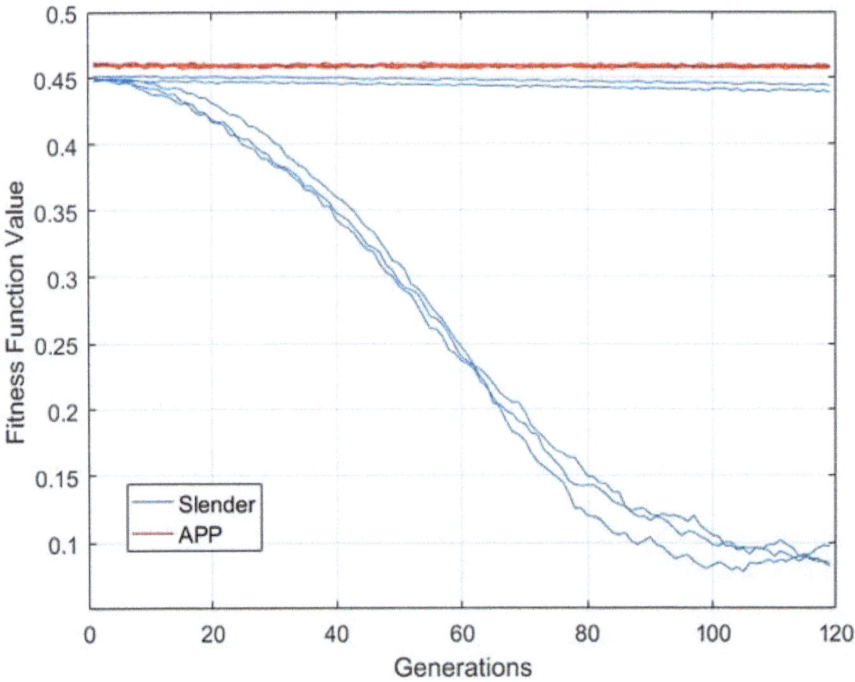

Fig. 11.13 Normalized fitness function values over 120 generations for attacks on Slender (blue) and the introduced APP (red). A lower value would indicate a better matching model

As the initial PUF model approaches 60% accuracy, the randomness of the g value becomes less effective in confusing the ES attack. This is why it is important to use a PUF circuit with high uniqueness values, which guarantees that initialized models would achieve around 50% accuracy. While it is unlikely that initial conditions can achieve such accuracy, utilizing a 3-XOR Arbiter PUF would ensure that such accuracy conditions would be impossible to initialize due to its high uniqueness.

Overall, the results showed that the introduced protocol is vastly superior to protocols like Slender when it comes to its resilience against machine learning attacks. Although the attacks failed to successfully model the Arbiter PUF circuit employed, we do not claim that the introduced protocol would be immune to such attacks when using a vulnerable PUF like the Arbiter PUF utilized in our experiment. Due to the low memory of the Arbiter PUF and its relatively low uniqueness, a dedicated attacker with access to high computational capacities might eventually generate an accurate model of the PUF with enough restarts. The Arbiter PUF was chosen for testing to have a better measure of the security gains of the protocol when utilizing more vulnerable PUFs.

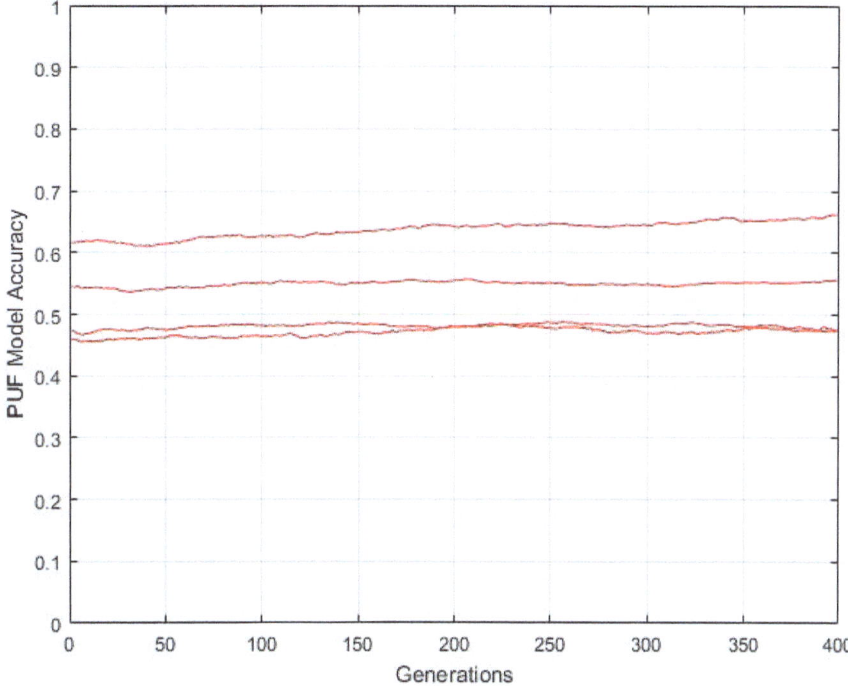

Fig. 11.14 Accuracy levels attained using a prolonged ES attack on APP with varying initial conditions

The experiment did clearly indicate the highly superior resilience the introduced protocol would offer. For reliable security, we would recommend the use of a 3-row XOR PUF or LS PUF when considering Arbiter-based PUFs. Any Strong PUFs with similar security can also be utilized by the protocol, as there is no restriction to the Strong PUF circuit besides the availability of a soft model.

To the best of our knowledge, the introduced protocol would be the first to protocol to show such high resilience against ES-based machine learning attacks while employing no cryptographic primitives and while also avoiding any limitations on the number of authentication attempts supported like the Lockdown protocol in [167]. The greater impact of the results is that lightweight authentication protocols utilizing strong PUFs are more than viable and are far from being a pipe dream as they can be resilient to machine learning attacks when properly implemented, even when using delay arbiter PUFs variants like the XOR PUF or LS PUF. The high effectiveness of machine learning attacks on strong PUFs has raised many questions about their usability, and the introduced Advanced PUF Protocol would renew confidence in the role strong PUFs can play in tackling emerging security threats.

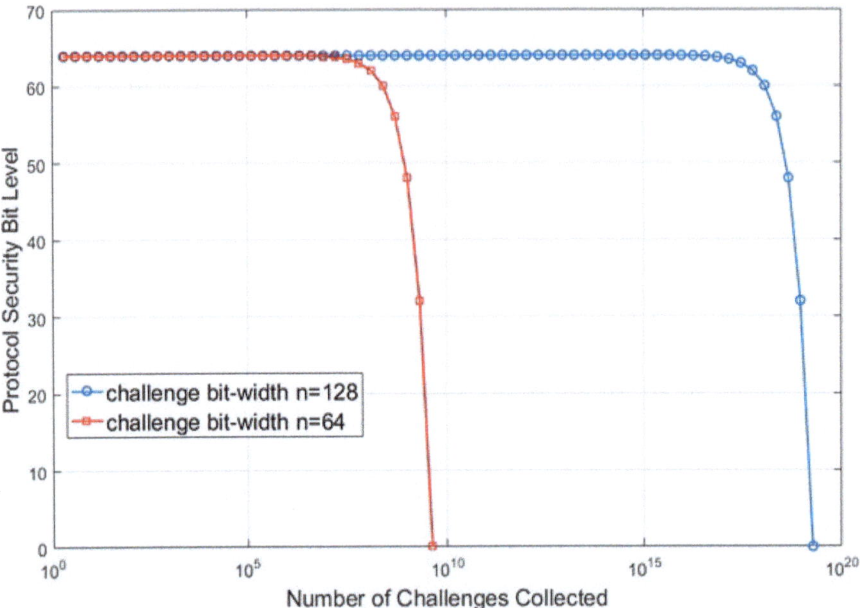

Fig. 11.15 The effect of the number of authentic pseudo-challenges collected by an adversary on the protocol bit-level security against random guess attacks

11.6.3   Exhaustive Challenge Collection Attack

By observing and storing the authentic exchange containing the pseudo-challenges $\{C1, C2, C3\}$, the attacker can produce an authentic response $\{C2, C3\}$ when presented with the same pseudo-challenge $C1$ by the verifier. An adversary would be able to use stored authentic exchanges to enhance their random guess attack efficiency. Figure 11.15 shows how the protocol security level against random guess attacks (y-axis) varies with respect to the number of authentic exchanges collected by the adversary (x-axis) for a challenge input bit-width of $n = 64$ (red) and $n = 128$ (blue). The figure shows that there would be no significant security loss until a significant portion of the challenges is collected.

The PUF is fully compromised when the attacker has collected an authentic challenge response $\{C2, C3\}$ for every possible challenge $C1$ that can be produced by the verifier. Since the prover controls half of the n bits of the challenge $C1$, this would require the observation of $2^{n/2}$ challenge exchanges and the storage of $2^{n/2} \times 3 \times n = 2^{0.5n + \log_2(3n)}$ bits making the attack completely infeasible to perform for a sufficiently large value of n as the protocol would guarantee a $0.5n$-bit security level against it. In our version of protocol, we utilize $n = 128$ to guarantee 64-bit security level.

While 64-bit cryptographic security is obsolete in modern-day technology, it is extremely secure against challenge collection attacks as the challenge collection cannot be parallelized or accelerated when it is limited to the speed of the associated device. Furthermore, the challenges observed need to be stored, which is infeasible. For instance, for $n = 128$, the adversary would need to store more than 800 exabytes of data.

11.7 Circuit Error Rate and Tuning

Due to the noisy outputs of PUF circuits, the authentication protocol would require employing some error tolerance value α. Such error tolerance would depend on the expected number of non-authentic challenges produced by an authentic node due to the natural circuit error rate of the PUF. A simple MATLAB binary search algorithm script was used along with the inverse binomial cumulative distribution function, also known as the quantile function, to find the minimum value of m that can maintain the desired random guess success probability while supporting the required error tolerance rate.

Figure 11.16 shows how the values of m and α increase to compensate for the PUF circuit error while maintaining a 64-bit security level. The fault tolerance value α is chosen to be 3% higher than the value of the expected error rate of non-authentic challenges E_{NA}. We note that the error rate of non-authentic challenges is higher than the PUF circuit error rate as multiple PUF challenge evaluations are involved in each authentication.

Table 11.6 highlights the tuning values of α and m at some key values of the PUF error rate. At each key point, the table shows the value of the PUF circuit error rate E_C, the expected non-authentic challenges rate in the protocol session E_{NA}, the tolerance rate t, the number of exchanges m that need to be utilized to maintain 64-bit security, and the number of allowed non-authentic exchanges α.

The table shows that the values of m and α remain manageable even when utilizing circuits with up to 8% error rate. The values can be tuned dynamically depending on the operating conditions and the aging effect of the circuit. We note though that when using the Advanced PUF Protocol for secret message exchange, higher error rate can be problematic even when the authenticity of the message can be confirmed. As such, it is advisable to use a highly reliable PUF in addition to error correction techniques when deploying the protocol for secret message exchange.

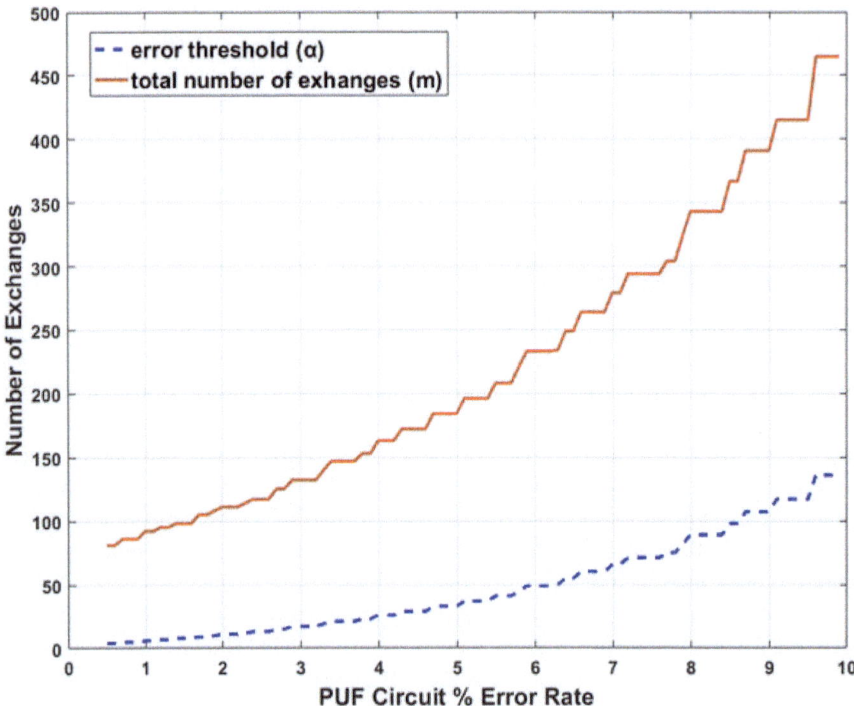

Fig. 11.16 The required growth of m and α to compensate for the PUF circuit error rate while maintaining a 64-bit level of security

Table 11.6 Number of exchanges required to maintain 64-bit security at key error rates

% PUF circuit error	% Faculty exchanges	% Fault tolerance	# Exchanges m	Fault tolerance threshold α
$E_c = 0.5\%$	$E_{NA} = 1.7\%$	$t = 4.7\%$	$m = 81$	$\alpha = 4$
$E_c = 4\%$	$E_{NA} = 12.6\%$	$t = 15.6\%$	$m = 26$	$\alpha = 4$
$E_c = 8\%$	$E_{NA} = 22.5\%$	$t = 25.5\%$	$m = 89$	$\alpha = 4$

11.8   Protocol System and FPGA Implementation

11.8.1   System Architecture

Figure 11.17 shows an illustration of how the protocol can be deployed on the device side and server side. On the server side, the protocol would be implemented in software. When the unconstrained device cannot be trusted with permanent access, the PUF model and the protocol logic can be stored on a remote trusted server, while unconstrained devices with temporary access can forward authentication requests to

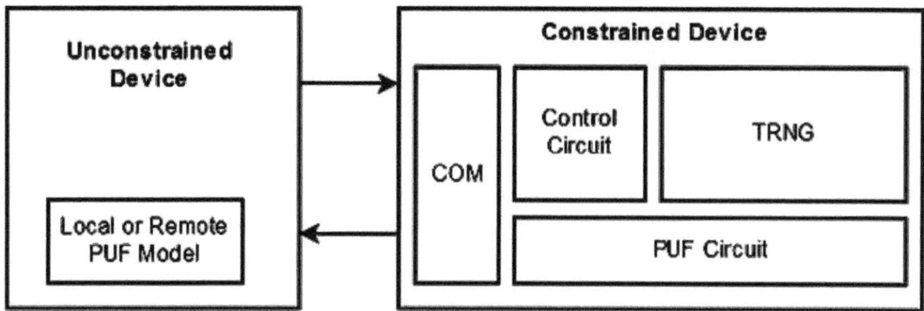

Fig. 11.17 Protocol system modules

the trusted remote server. This would allow system administrators to revoke the access rights by simply denying devices access to the remote model and logic.

On the device side, the protocol would require the implementation of a strong Arbiter PUF circuit, a TRNG, and a control circuit to handle the authentication logic. The components and modules required are very similar to those utilized by the Slender PUF Protocol [109]. However, the Advanced PUF Protocol (APP) utilizes a more complex control logic for the authentication process.

11.8.2   FPGA Implementation

While the authentication logic can be handled by a microcontroller, it can still be easily implemented through hardware with a small implementation area overhead. To verify the control logic's low implementation area overhead and the functionality of the protocol, an FPGA implementation was performed. The protocol modules were implemented on a Xilinx Artix-7 FPGA. A 131-bit Arbiter PUF was used in our implementation. The implemented PUF uses LUTs as switching elements. We note that the original delay arbiter PUF would not be suitable for FPGA implementation as it would yield very low uniqueness. Nonetheless, as the purpose of the implementation is to verify the functionality of the protocol and its low control logic overhead, we sufficed with the use of the Arbiter PUF.

The control circuit would handle all authentication and challenge transformation logic. The implemented version of the protocol requires m=163 number of rounds for each authentication while allowing for $\alpha = 29$ non-authentic challenge exchanges. This allows the use of PUF circuits with up to 4% error rate.

There are various TRNGs that would be suitable for our protocol FPGA verification [129, 148]. However, we simply utilized a small PRNG that would suffice for the verification purpose of our implementation. We note, though, that the protocol introduced is meant to be implemented in ASIC technology for a lower power and a lower implementation area overhead and, as such, would utilize analog-

Table 11.7 Number of LUTs
and registers used in the
FPGA implementation

Implemented module	LUT	Registers
PUF circuit	262	1
Control circuit	59	35

based TRNGs, which are suitable for constrained devices like those introduced
in [27, 83] or [29]. A Universal Asynchronous Receiver-Transmitter module was
implemented to handle the FPGA-PC communication. Table 11.7 shows the number
of LUTs and registers consumed by each of the control circuit module and the PUF
circuit module in our implementation. The results show that the required increase in
the implementation area imposed by the control circuit would be marginal when
compared to the size of the PUF circuit. A total of 59 LUTs and 35 registers
would suffice to implement the control logic of the Advanced PUF Protocol when a
microcontroller is not available for a constrained device.

11.9   Related Work Comparison

There have been several attempts on lightweight PUF-based authentication protocols that avoid the use of cryptographic functions like encryption or hashing.
However, the protocols introduced so far suffer from various disadvantages or
vulnerabilities when compared to the Advanced PUF Protocol. The authors in
[34] did a comprehensive survey on various PUF-based protocols. We compare
our work with the most successful lightweight protocols from their work, along
with recently introduced protocols that include the Lockdown protocol [167] and
challenge-obfuscation [51] protocol.

11.9.1   Lockdown Protocol

Lockdown protocol [167] limits the number of possible authentication attempts a
device can perform. This ensures that a PUF would never reveal enough challenge-
response pairs to allow an adversary to model the PUF circuit. The protocol employs
a counter at the server side to keep track of the number of authentication attempts
performed. Such a solution is very simplistic and ineffective as the protocol fails to
address the main challenge associated with it, which would be securing the protocol
against Denial of Service (DoS) attacks that aim to exhaust the supported number
of authentications.

The authors did recognize the DoS threat and suggested that the server involved
in the authentication should employ a DoS detection mechanism. However, no
explanations or details of such mechanism nor its effectiveness have been provided.
DoS attacks are very easy to perform, and there is no solid security measure that

can completely stop these attacks. By simply jamming the communication channel or impersonating a device, an adversary can quickly exhaust the limited number of supported authentications. The lockdown protocol would be as such vulnerable to DoS attacks that exhaust its limited number of supported authentications rendering the PUF device unusable.

11.9.2 Slender Protocol

The Slender protocol introduced in [109] is similar to the Advanced PUF-based Protocol when it comes to the building blocks used: a TRNG, a PUF circuit, and a small control circuit. Slender protocol hides the response of the PUF circuit in plain sight by concatenating the PUF responses with a random bit stream generated by a TRNG and then logically rotating the string by random number to hide the index of the authentic response substring. The server receiving the full string would then perform a substring matching algorithm to determine whether the authentic substring is part of the device response.

The protocol however is vulnerable to a man-in-the-middle attacks that can expose the location of the substring index or corrupt its secret key exchange. Furthermore, it fails to offer mutual authentication as the substring matching algorithm is too computationally expensive for constrained devices. In [10] it was shown that ES algorithm can successfully produce an accurate model of the PUF circuit employed. In this chapter, we introduce a Man-in-the-Middle attack (MITM) that can further reduce the security of the protocol by exposing the location of the authentic index in a linear number of attempts. To the best of our knowledge, this attack on the Slender protocol has not been revealed in prior literature.

Figure 11.18 shows how a man in the middle can substitute several bits in the prover response with their inverted values. The substitutions target two portions of the string that are $(L_a - q)$ bits apart, where L_a is the length of the authentic substring and q the number of bit-flips introduced. The adversary introduces just enough bit-flips to ensure that the error tolerance of the protocol is exceeded. If the substituted bits end up being part of the authentic substring, then the device authentication would fail.

When the adversary observes an authentication failure after performing this substitution attack, they can conclude that the substituted bits were part of the

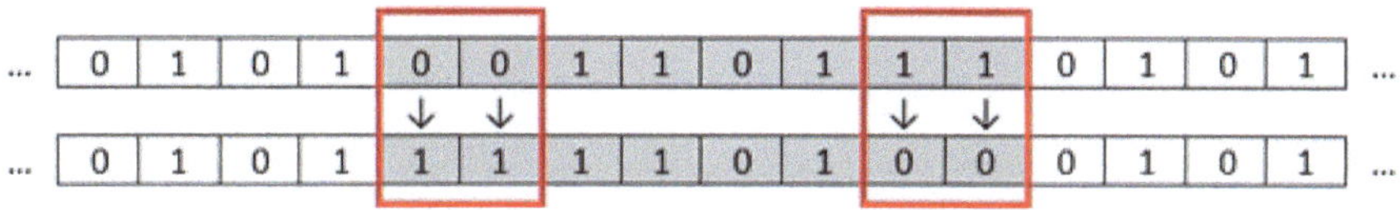

Fig. 11.18 MITM substitution attack performed on the Slender protocol. The attacker targets two blocks of bits that are La-q bits apart in an attempt to cause an authentication failure

authentic substring. Otherwise, the authentication would have succeeded. Since the two substituted portions are $(L_a - q)$ bits apart, the adversary can also conclude that the substitutions occurred at the boundaries of the authentic substring. This would reveal the location of the authentic substring with a high accuracy and allow the adversary easier access to the challenge-response pairs. The probability P_S for such an attack to succeed would be equal to the probability of choosing a random index that happens to be the index of the authentic substring starts. Given that L_S is the length of the full string, P_S would be defined as

$$P_S = \frac{1}{L_S}. \tag{11.3}$$

We can see that the protocol can only offer linear security against this attack as the probability of success is proportional to $1/L_S$. To achieve a bit-level security of 64-bits, L_S would need to have a value of 2^{64} that is infeasible. In the version of the protocol that is used for secret key exchange, the index of the authentic substring would be the secret message exchange. This secret key exchange is vulnerable to an MITM attack where the adversary rotates the transmitted bit string and alters the index value while preserving the authenticity of the message.

11.9.3   Other Protocols and Summary of Comparisons

Another recently introduced lightweight authentication method is the Obfuscated challenge-response [51]. However, The Obfuscated challenge-response protocol does not offer mutual authentication and would require a large circuit of XOR-PUFs to achieve a solid security level. In [169], the Noise Bifurcation Protocol showed increased resistance against modeling attacks. However, it does not claim a guaranteed level of security, and it is also unable to offer mutual authentication. Table 11.8 compares the Advanced PUF Protocol (APP) with the other mentioned lightweight PUF-based protocols.

The introduced protocol shows several advantages over the other lightweight protocols. Neither slender [109] nor noise bifurcation [169] nor obfuscated challenge-response protocols offer mutual authentication. Moreover, all three protocols suffer from increased vulnerability to modeling attacks when compared to the Advanced

Table 11.8 Comparison between APP and other PUF-based protocols

Protocol	TRNG required	Number of auth.	Modeling robust	Mutual auth.	Secret mess
Slender	✓	∞	~	✗	~
Noise bifur.	✓	∞	~	✗	✗
Lockdown llb	✓	d	✓	✓	✗
Obfuscated Ch.	✓	∞	~	✗	✗
APP (this work)	✓	∞	✓	✓	✓

PUF-based Protocol. While lockdown IIb [167] does provide mutual authentication, it can only support a limited number of authentications, which makes it also vulnerable to DoS attacks. The introduced Advanced PUF Protocol (APP) has none of these shortcomings, and its high resilience against machine learning attacks like ES would put it in a different league when compared to known protocols that do not utilize cryptographic primitives. Furthermore, the protocol would be the first to allow for authenticated secret message exchange between all parties.

References

1. Threat landscape for industrial automation systems in h2 2017, Aug 2018.
2. Sonia Akter, Kasem Khalil, and Magdy Bayoumi. Hardware security in the Internet of Things: A survey. In *2023 IEEE 36th International System-on-Chip Conference (SOCC)*, pages 1–6. IEEE, 2023.
3. Sonia Akter, Kasem Khalil, and Magdy Bayoumi. A survey on hardware security: Current trends and challenges. *IEEE Access*, 2023.
4. Sara Alahmadi, Kasem Khalil, Haytham Idriss, and Magdy Bayoumi. Fortifying strong PUFs: A modeling attack-resilient approach using weak PUF for IoT device security. In *2024 IEEE International Symposium on Circuits and Systems (ISCAS)*, pages 1–5. IEEE, 2024.
5. Usman Ali and Omer Khan. Connoc: A practical timing channel attack on network-on-chip hardware in a multicore processor. In *2021 IEEE International Symposium on Hardware Oriented Security and Trust (HOST)*, pages 192–202. IEEE, 2021.
6. Felipe Almeida, Malik Imran, Jaan Raik, and Samuel Pagliarini. Ransomware attack as hardware trojan: A feasibility and demonstration study. *IEEE Access*, 10:44827–44839, 2022.
7. Ahmad O Aseeri, Yu Zhuang, and Mohammed Saeed Alkatheiri. A machine learning-based security vulnerability study on XOR PUFs for resource-constraint Internet of Things. In *2018 IEEE International Congress on Internet of Things (ICIOT)*, pages 49–56. IEEE, 2018.
8. Elaine Barker, Don Johnson, and Miles Smid. *Recommendation for pair-wise key establishment schemes using discrete logarithm cryptography*. National Institute of Standards and Technology, 2006.
9. Georg T Becker. On the pitfalls of using arbiter-PUFs as building blocks. *IEEE Transactions on Computer-Aided Design of Integrated Circuits and Systems*, 34(8):1295–1307, 2015.
10. Georg T Becker. On the pitfalls of using arbiter-PUFs as building blocks. *IEEE Transactions on Computer-Aided Design of Integrated Circuits and Systems*, 34(8):1295–1307, 2015.
11. Georg T Becker, Alexander Wild, and Tim Güneysu. Security analysis of index-based syndrome coding for PUF-based key generation. In *2015 IEEE International Symposium on Hardware Oriented Security and Trust (HOST)*, pages 20–25. IEEE, 2015.
12. Matthew Becker and Ahmed Desoky. A study of the DVD content scrambling system (CSS) algorithm. In *Proceedings of the Fourth IEEE International Symposium on Signal Processing and Information Technology, 2004.*, pages 353–356. IEEE, 2004.
13. Mihir Bellare, David Pointcheval, and Phillip Rogaway. Authenticated key exchange secure against dictionary attacks. In *International conference on the theory and applications of cryptographic techniques*, pages 139–155. Springer, 2000.

14. Mudit Bhargava and Ken Mai. An efficient reliable PUF-based cryptographic key generator in 65 nm CMOS. In *2014 Design, Automation & Test in Europe Conference & Exhibition (DATE)*, pages 1–6. IEEE, 2014.

15. Eli Biham and Adi Shamir. Differential fault analysis of secret key cryptosystems. In *Annual international cryptology conference*, pages 513–525. Springer, 1997.

16. Dennis Blair, John Huntsman, C Barrett, W Lynn, S Gorton, D Wince-Smith, and M Young. update to the IP commission report. *The Commission on the Theft of American Intellectual Property by The National Bureau of Asian Research*, 2017.

17. Andrey Bogdanov, Lars R Knudsen, Gregor Leander, Christof Paar, Axel Poschmann, Matthew JB Robshaw, Yannick Seurin, and Charlotte Vikkelsoe. Present: An ultra-lightweight block cipher. In *International Workshop on Cryptographic Hardware and Embedded Systems*, pages 450–466. Springer, 2007.

18. Christoph Böhm and Maximilian Hofer. *Physical unclonable functions in theory and practice*. Springer Science & Business Media, 2012.

19. Mark Bohr. The new era of scaling in an SOC world. In *2009 IEEE International Solid-State Circuits Conference-Digest of Technical Papers*, pages 23–28. IEEE, 2009.

20. Victor Boyko, Philip MacKenzie, and Sarvar Patel. Provably secure password-authenticated key exchange using Diffie–Hellman. In *International Conference on the Theory and Applications of Cryptographic Techniques*, pages 156–171. Springer, 2000.

21. Eric Brier, Christophe Clavier, and Francis Olivier. Correlation power analysis with a leakage model. In *International workshop on cryptographic hardware and embedded systems*, pages 16–29. Springer, 2004.

22. Kunbei Cai, Md Hafizul Islam Chowdhuryy, Zhenkai Zhang, and Fan Yao. Seeds of seed: NMT-Stroke: Diverting neural machine translation through hardware-based faults. In *2021 International Symposium on Secure and Private Execution Environment Design (SEED)*, pages 76–82. IEEE, 2021.

23. Kunbei Cai, Zhenkai Zhang, and Fan Yao. On the feasibility of training-time trojan attacks through hardware-based faults in memory. In *2022 IEEE International Symposium on Hardware Oriented Security and Trust (HOST)*, pages 133–136. IEEE, 2022.

24. Francois Cayre and Patrick Bas. Kerckhoffs-based embedding security classes for WOA data hiding. *IEEE Transactions on Information Forensics and Security*, 3(1):1–15, 2008.

25. Rajat Subhra Chakraborty and Swarup Bhunia. Harpoon: An obfuscation-based SOC design methodology for hardware protection. *IEEE Transactions on Computer-Aided Design of Integrated Circuits and Systems*, 28(10):1493–1502, 2009.

26. Qian Chen and Robert A Bridges. Automated behavioral analysis of malware: A case study of WannaCry ransomware. In *2017 16th IEEE International Conference on machine learning and applications (ICMLA)*, pages 454–460. IEEE, 2017.

27. Wei Chen, Wenyi Che, Zhongyu Bi, Jing Wang, Na Yan, Xi Tan, Junyu Wang, Hao Min, and Jie Tan. A 1.04 μw truly random number generator for gen2 RFID tag. In *2009 IEEE Asian Solid-State Circuits Conference*, pages 117–120. IEEE, 2009.

28. Xinyun Chen, Chang Liu, Bo Li, Kimberly Lu, and Dawn Song. Targeted backdoor attacks on deep learning systems using data poisoning. *arXiv preprint arXiv:1712.05526*, 2017.

29. Zhengyu Chen, Wenbo Yin, and Kevin Boyle. True random number generator in 0.35/μm for RFID applications. In *2015 IEEE International Conference on Electron Devices and Solid-State Circuits (EDSSC)*, pages 625–628. IEEE, 2015.

30. Hung-Yu Chien. Sasi: A new ultralightweight RFID authentication protocol providing strong authentication and strong integrity. *IEEE transactions on dependable and secure computing*, 4(4):337–340, 2007.

31. Joan Daemen and Vincent Rijmen. *The design of Rijndael: AES-the advanced encryption standard*. Springer Science & Business Media, 2013.

32. Christophe De Cannière. Trivium: A stream cipher construction inspired by block cipher design principles. In *International Conference on Information Security*, pages 171–186. Springer, 2006.

33. Jeroen Delvaux, Roel Peeters, Dawu Gu, and Ingrid Verbauwhede. A survey on lightweight entity authentication with strong PUFs. *ACM Computing Surveys (CSUR)*, 48(2):1–42, 2015.

34. Jeroen Delvaux, Roel Peeters, Dawu Gu, and Ingrid Verbauwhede. A survey on lightweight entity authentication with strong PUFs. *ACM Computing Surveys (CSUR)*, 48(2):26, 2015.

35. Jeroen Delvaux and Ingrid Verbauwhede. Side channel modeling attacks on 65 nm arbiter PUFs exploiting CMOS device noise. In *2013 IEEE International Symposium on Hardware-Oriented Security and Trust (HOST)*, pages 137–142. IEEE, 2013.

36. Jeroen Delvaux and Ingrid Verbauwhede. Attacking PUF-based pattern matching key generators via helper data manipulation. In *Cryptographers' Track at the RSA Conference*, pages 106–131. Springer, 2014.

37. Bappaditya Dey, Kasem Khalil, Ashok Kumar, and Magdy Bayoumi. A novel design gate based low-cost configurable ro PUF using reversible logic. In *2019 IEEE 62nd International Midwest Symposium on Circuits and Systems (MWSCAS)*, pages 211–214. IEEE, 2019.

38. E. Dubrova, O. Näslund, B. Degen, A. Gawell, and Y. Yu. CRC-PUF: A machine learning attack resistant lightweight PUF construction. In *2019 IEEE European Symposium on Security and Privacy Workshops (EuroS PW)*, pages 264–271, 2019.

39. D Eastlake 3rd and Tony Hansen. US secure hash algorithms (SHA and SHA-based HMAC and HKDF). Technical report, 2011.

40. D Eastlake 3rd and Paul Jones. US secure hash algorithm 1 (SHA1). Technical report, 2001.

41. Taher ElGamal. A public key cryptosystem and a signature scheme based on discrete logarithms. *IEEE transactions on information theory*, 31(4):469–472, 1985.

42. Samuel Ellicott, Michael Kines, Waleed Khalil, Yu Qi, Abdullah Kurtoglu, and Hossein Miri Lavasani. Analog-inspired hardware security: A low-energy solution for IoT trusted communications. In *2021 IEEE 34th International System-on-Chip Conference (SOCC)*, pages 200–205. IEEE, 2021.

43. Mohamed Elshamy, Giorgio Di Natale, Antonios Pavlidis, Marie-Minerve Louërat, and Haralampos-G Stratigopoulos. Hardware trojan attacks in analog/mixed-signal ICS via the test access mechanism. In *2020 IEEE European Test Symposium (ETS)*, pages 1–6. IEEE, 2020.

44. Mohammad Eslami, Tara Ghasempouri, and Samuel Pagliarini. Reusing verification assertions as security checkers for hardware trojan detection. *arXiv preprint arXiv:2201.01130*, 2022.

45. Xuejun Fan. Wafer level packaging (WLP): fan-in, fan-out and three-dimensional integration. In *2010 11th International Thermal, Mechanical & Multi-Physics Simulation, and Experiments in Microelectronics and Microsystems (EuroSimE)*, pages 1–7. IEEE, 2010.

46. Klaus Finkenzeller. *RFID handbook: fundamentals and applications in contactless smart cards, radio frequency identification and near-field communication*. John wiley & sons, 2010.

47. Patrick Gallagher. Digital signature standard (DSS). *Federal Information Processing Standards Publications, volume FIPS*, pages 186–3, 2013.

48. Fatemeh Ganji, Shahin Tajik, and Jean-Pierre Seifert. Why attackers win: on the learnability of XOR arbiter PUFs. In *International Conference on Trust and Trustworthy Computing*, pages 22–39. Springer, 2015.

49. Fatemeh Ganji, Shahin Tajik, and Jean-Pierre Seifert. Why attackers win: on the learnability of XOR arbiter PUFs. In *International Conference on Trust and Trustworthy Computing*, pages 22–39. Springer, 2015.

50. Yansong Gao, Gefei Li, Hua Ma, Said F Al-Sarawi, Omid Kavehei, Derek Abbott, and Damith C Ranasinghe. Obfuscated challenge-response: A secure lightweight authentication mechanism for PUF-based pervasive devices. In *2016 IEEE International Conference on Pervasive Computing and Communication Workshops (PerCom Workshops)*, pages 1–6. IEEE, 2016.

51. Yansong Gao, Gefei Li, Hua Ma, Said F Al-Sarawi, Omid Kavehei, Derek Abbott, and Damith C Ranasinghe. Obfuscated challenge-response: A secure lightweight authentication mechanism for PUF-based pervasive devices. In *2016 IEEE International Conference on Pervasive Computing and Communication Workshops (PerCom Workshops)*, pages 1–6. IEEE, 2016.

52. Yansong Gao, Hua Ma, Geifei Li, Shaza Zeitouni, Said F Al-Sarawi, Derek Abbott, Ahmad-Reza Sadeghi, and Damith C Ranasinghe. Exploiting PUF models for error free response generation. *arXiv preprint arXiv:1701.08241*, 2017.

53. Blaise Gassend, Dwaine Clarke, Marten Van Dijk, and Srinivas Devadas. Controlled physical random functions. In *18th Annual Computer Security Applications Conference, 2002. Proceedings.*, pages 149–160. IEEE, 2002.

54. Blaise Gassend, Dwaine Clarke, Marten Van Dijk, and Srinivas Devadas. Silicon physical random functions. In *Proceedings of the 9th ACM conference on Computer and communications security*, pages 148–160. ACM, 2002.

55. Maël Gay, Tobias Paxian, Devanshi Upadhyaya, Bernd Becker, and Ilia Polian. Hardware-oriented algebraic fault attack framework with multiple fault injection support. In *2019 Workshop on Fault Diagnosis and Tolerance in Cryptography (FDTC)*, pages 25–32. IEEE, 2019.

56. Daniel Genkin, Adi Shamir, and Eran Tromer. RSA key extraction via low-bandwidth acoustic cryptanalysis. In *Annual cryptology conference*, pages 444–461. Springer, 2014.

57. John Greenough. How the 'Internet of Things' will affect the world, Jan 2016.

58. Amit Grover and Hal Berghel. A survey of RFID deployment and security issues. *Journal of information processing systems*, 7(4):561–580, 2011.

59. Chongyan Gu, Chip-Hong Chang, Weiqiang Liu, Shichao Yu, Yale Wang, and Máire O'Neill. A modeling attack resistant deception technique for securing lightweight-PUF-based authentication. *IEEE Transactions on Computer-Aided Design of Integrated Circuits and Systems*, 40(6):1183–1196, 2020.

60. Jorge Guajardo, Sandeep S Kumar, Geert-Jan Schrijen, and Pim Tuyls. FPGA intrinsic PUFs and their use for ip protection. In *International workshop on cryptographic hardware and embedded systems*, pages 63–80. Springer, 2007.

61. Nikolaus Hansen. The CMA evolution strategy: a comparing review. In *Towards a new evolutionary computation*, pages 75–102. Springer, 2006.

62. Zhangqing He, Meilin Wan, Jie Deng, Chuang Bai, and Kui Dai. A reliable strong PUF based on switched-capacitor circuit. *IEEE Transactions on Very Large Scale Integration (VLSI) Systems*, 26(6):1073–1083, 2018.

63. Charles Herder, Meng-Day Yu, Farinaz Koushanfar, and Srinivas Devadas. Physical unclonable functions and applications: A tutorial. *Proceedings of the IEEE*, 102(8):1126–1141, 2014.

64. Alex Hern. Fitness tracking app strava gives away location of secret us army bases, Jan 2018.

65. Benjamin Hettwer, Daniel Fennes, Sebastien Leger, Jan Richter-Brockmann, Stefan Gehrer, and Tim Güneysu. Deep learning multi-channel fusion attack against side-channel protected hardware. In *2020 57th ACM/IEEE Design Automation Conference (DAC)*, pages 1–6. IEEE, 2020.

66. Matthew Hicks, Murph Finnicum, Samuel T King, Milo MK Martin, and Jonathan M Smith. Overcoming an untrusted computing base: Detecting and removing malicious hardware automatically. In *2010 IEEE symposium on security and privacy*, pages 159–172. IEEE, 2010.

67. Yohei Hori, Takahiro Yoshida, Toshihiro Katashita, and Akashi Satoh. Quantitative and statistical performance evaluation of arbiter physical unclonable functions on FPGAs. In *2010 International Conference on Reconfigurable Computing and FPGAs*, pages 298–303. IEEE, 2010.

68. Wei Hu, Chip-Hong Chang, Anirban Sengupta, Swarup Bhunia, Ryan Kastner, and Hai Li. An overview of hardware security and trust: Threats, countermeasures, and design tools. *IEEE Transactions on Computer-Aided Design of Integrated Circuits and Systems*, 40(6):1010–1038, 2020.

69. Haytham Idriss, Tarek Idriss, and Magdy Bayoumi. A highly reliable dual-arbiter PUF for lightweight authentication protocols. In *2017 IEEE International Conference on RFID Technology & Application (RFID-TA)*, pages 248–253. IEEE, 2017.

70. Tarek Idriss, Haytham Idriss, and Magdy Bayoumi. A PUF-based paradigm for IoT security. In *2016 IEEE 3rd World Forum on Internet of Things (WF-IoT)*, pages 700–705. IEEE, 2016.

71. Tarek A. Idriss, Haytham A. Idriss, and Magdy A. Bayoumi. A lightweight PUF-based authentication protocol using secret pattern recognition for constrained IoT devices. *IEEE Access*, 9:80546–80558, 2021.

72. Nikolay Ivanov and Qiben Yan. Ethclipper: A clipboard meddling attack on hardware wallets with address verification evasion. In *2021 IEEE Conference on Communications and Network Security (CNS)*, pages 191–199. IEEE, 2021.

73. Ayush Jain and Ujjwal Guin. A novel tampering attack on AES cores with hardware trojans. In *2020 IEEE International Test Conference in Asia (ITC-Asia)*, pages 77–82. IEEE, 2020.

74. Naghmeh Karimi, Jean-Luc Danger, Florent Lozac'h, and Sylvain Guilley. Predictive aging of reliability of two delay PUFs. In *International Conference on Security, Privacy, and Applied Cryptography Engineering*, pages 213–232. Springer, 2016.

75. Ramesh Karri, Jeyavijayan Rajendran, Kurt Rosenfeld, and Mohammad Tehranipoor. Trustworthy hardware: Identifying and classifying hardware trojans. *Computer*, 43(10):39–46, 2010.

76. Athanasios T Karygiannis, Bernard Eydt, Greg Barber, Lynn Bunn, and Ted Phillips. SP 800-98. guidelines for securing radio frequency identification (RFID) systems, 2007.

77. Kasem Khalil, Ahmed Abdelgawad, and Magdy Bayoumi. Intelligent resource discovery approach for the Internet of Things. In *2021 IEEE 7th World Forum on Internet of Things (WF-IoT)*, pages 264–269. IEEE, 2021.

78. Kasem Khalil, Khalid Elgazzar, Ahmed Abdelgawad, and Magdy Bayoumi. A security approach for CoAP-based Internet of Things resource discovery. In *2020 IEEE 6th World Forum on Internet of Things (WF-IoT)*, pages 1–6. IEEE, 2020.

79. Kasem Khalil, Khalid Elgazzar, and Magdy Bayoumi. A comparative analysis on resource discovery protocols for the Internet of Things. In *2018 IEEE Global Communications Conference (GLOBECOM)*, pages 1–7. IEEE, 2018.

80. Kasem Khalil, Khalid Elgazzar, Mohamed Seliem, and Magdy Bayoumi. Resource discovery techniques in the Internet of Things: a review. *Internet of Things*, 12:100293, 2020.

81. Kasem Khalil, Tamador Mohaidat, Mahmoud Darwich, Ashok Kumar, and Magdy Bayoumi. An efficient hardware design of CoAP protocol for the internet of things. In *2024 IEEE 17th Dallas Circuits and Systems Conference (DCAS)*, pages 1–5. IEEE, 2024.

82. Mohamed Amine Khelif, Jordane Lorandel, Olivier Romain, Matthieu Regnery, Denis Baheux, and Guillaume Barbu. Toward a hardware man-in-the-middle attack on PCIe bus for smart data replay. In *2019 22nd Euromicro Conference on Digital System Design (DSD)*, pages 230–237. IEEE, 2019.

83. Minseo Kim, Unsoo Ha, Kyuho Jason Lee, Yongsu Lee, and Hoi-Jun Yoo. A 82-nw chaotic map true random number generator based on a sub-ranging SAR ADC. *IEEE Journal of Solid-State Circuits*, 52(7):1953–1965, 2017.

84. Yoongu Kim, Ross Daly, Jeremie Kim, Chris Fallin, Ji Hye Lee, Donghyuk Lee, Chris Wilkerson, Konrad Lai, and Onur Mutlu. Flipping bits in memory without accessing them: An experimental study of dram disturbance errors. *ACM SIGARCH Computer Architecture News*, 42(3):361–372, 2014.

85. Paul Kocher, Joshua Jaffe, and Benjamin Jun. Differential power analysis. In *Annual international cryptology conference*, pages 388–397. Springer, 1999.

86. S. T. C. Konigsmark, D. Chen, and M. D. F. Wong. Polypuf: Physically secure self-divergence. *IEEE Transactions on Computer-Aided Design of Integrated Circuits and Systems*, 35(7):1053–1066, 2016.

87. ST Konigsmark, Leslie K Hwang, Deming Chen, and Martin DF Wong. System-of-PUFs: multilevel security for embedded systems. In *Proceedings of the 2014 International Conference on Hardware/Software Codesign and System Synthesis*, page 27. ACM, 2014.

88. Farinaz Koushanfar, Saverio Fazzari, Carl McCants, William Bryson, Matthew Sale, Peilin Song, and Miodrag Potkonjak. Can EDA combat the rise of electronic counterfeiting? In *Proceedings of the 49th Annual Design Automation Conference*, pages 133–138, 2012.

89. Vedika J Kulkarni, R Manju, Ruchika Gupta, John Jose, and Sukumar Nandi. Packet header attack by hardware trojan in NOC based TCMP and its impact analysis. In *Proceedings of the 15th IEEE/ACM International Symposium on Networks-on-Chip*, pages 21–28, 2021.

90. Man-Hsuan Kuo, Chun-Ming Hu, and Kuen-Jong Lee. Time-related hardware trojan attacks on processor cores. In *2019 IEEE International Test Conference in Asia (ITC-Asia)*, pages 43–48. IEEE, 2019.

91. Yingjie Lao, Bo Yuan, Chris H Kim, and Keshab K Parhi. Reliable PUF-based local authentication with self-correction. *IEEE Transactions on Computer-Aided Design of Integrated Circuits and Systems*, 36(2):201–213, 2016.

92. Jae W Lee, Daihyun Lim, Blaise Gassend, G Edward Suh, Marten Van Dijk, and Srinivas Devadas. A technique to build a secret key in integrated circuits for identification and authentication applications. In *2004 Symposium on VLSI Circuits. Digest of Technical Papers (IEEE Cat. No. 04CH37525)*, pages 176–179. IEEE, 2004.

93. Jae W Lee, Daihyun Lim, Blaise Gassend, G Edward Suh, Marten Van Dijk, and Srinivas Devadas. A technique to build a secret key in integrated circuits for identification and authentication applications. In *2004 Symposium on VLSI Circuits. Digest of Technical Papers (IEEE Cat. No. 04CH37525)*, pages 176–179. IEEE, 2004.

94. Seon-Kyoo Lee, Young-Hun Seo, Hong-June Park, and Jae-Yoon Sim. A 1 GHz ADPLL with a 1.25 ps minimum-resolution sub-exponent TDC in $0.18\,\mu$ m CMOS. *IEEE Journal of Solid-State Circuits*, 45(12):2874–2881, 2010.

95. Hong Li, YongHui Chen, and ZhangQing He. The survey of RFID attacks and defenses. In *2012 8th International Conference on Wireless Communications, Networking and Mobile Computing*, pages 1–4. ieee, 2012.

96. Lang Lin, Dan Holcomb, Dilip Kumar Krishnappa, Prasad Shabadi, and Wayne Burleson. Low-power sub-threshold design of secure physical unclonable functions. In *Proceedings of the 16th ACM/IEEE international symposium on Low power electronics and design*, pages 43–48, 2010.

97. Dongsheng Liu, Zilong Liu, Zhenqiang Yong, Xuecheng Zou, and Jian Cheng. Design and implementation of an ECC-based digital baseband controller for RFID tag chip. *IEEE Transactions on Industrial Electronics*, 62(7):4365–4373, 2015.

98. Dongsheng Liu, Zilong Liu, Zhenqiang Yong, Xuecheng Zou, and Jian Cheng. Design and implementation of an ECC-based digital baseband controller for RFID tag chip. *IEEE Transactions on Industrial Electronics*, 62(7):4365–4373, 2015.

99. Muqing Liu, Chen Zhou, Qianying Tang, Keshab K Parhi, and Chris H Kim. A data remanence based approach to generate 100% stable keys from an SRAM physical unclonable function. In *2017 IEEE/ACM International Symposium on Low Power Electronics and Design (ISLPED)*, pages 1–6. IEEE, 2017.

100. Yingqi Liu, Shiqing Ma, Yousra Aafer, Wen-Chuan Lee, Juan Zhai, Weihang Wang, and Xiangyu Zhang. Trojaning attack on neural networks. 2017.

101. Lifars Llc. FDA recalls half a million pacemakers due to hacking fears, Sep 2017.

102. Roel Maes, Pim Tuyls, and Ingrid Verbauwhede. A soft decision helper data algorithm for SRAM PUFs. In *2009 IEEE international symposium on information theory*, pages 2101–2105. IEEE, 2009.

103. Abhranil Maiti, Jeff Casarona, Luke McHale, and Patrick Schaumont. A large scale characterization of ro-PUF. In *2010 IEEE International Symposium on Hardware-Oriented Security and Trust (HOST)*, pages 94–99. IEEE, 2010.

104. M. Majzoobi, M. Rostami, F. Koushanfar, D. S. Wallach, and S. Devadas. Slender PUF protocol: A lightweight, robust, and secure authentication by substring matching. In *2012 IEEE Symposium on Security and Privacy Workshops*, pages 33–44, 2012.

105. Mehrdad Majzoobi, Farinaz Koushanfar, and Miodrag Potkonjak. Lightweight secure PUFs. In *2008 IEEE/ACM International Conference on Computer-Aided Design*, pages 670–673. IEEE, 2008.

106. Mehrdad Majzoobi, Farinaz Koushanfar, and Miodrag Potkonjak. Lightweight secure PUFs. In *Proceedings of the 2008 IEEE/ACM International Conference on Computer-Aided Design*, pages 670–673. IEEE Press, 2008.

107. Mehrdad Majzoobi, Farinaz Koushanfar, and Miodrag Potkonjak. Testing techniques for hardware security. In *2008 IEEE International Test Conference*, pages 1–10. IEEE, 2008.

108. Mehrdad Majzoobi, Masoud Rostami, Farinaz Koushanfar, Dan S Wallach, and Srinivas Devadas. Slender PUF protocol: A lightweight, robust, and secure authentication by substring matching. In *2012 IEEE Symposium on Security and Privacy Workshops*, pages 33–44. IEEE, 2012.

109. Mehrdad Majzoobi, Masoud Rostami, Farinaz Koushanfar, Dan S Wallach, and Srinivas Devadas. Slender PUF protocol: A lightweight, robust, and secure authentication by substring matching. In *2012 IEEE Symposium on Security and Privacy Workshops*, pages 33–44. IEEE, 2012.

110. Priyanka Mall, Ruhul Amin, Ashok Kumar Das, Mark T Leung, and Kim-Kwang Raymond Choo. PUF-based authentication and key agreement protocols for IoT, WSNs, and smart grids: a comprehensive survey. *IEEE Internet of Things Journal*, 9(11):8205–8228, 2022.

111. The Mathworks, Inc., Natick, Massachusetts. *MATLAB version 9.3.0.713579 (R2017b)*, 2017.

112. Dominik Merli, Johann Heyszl, Benedikt Heinz, Dieter Schuster, Frederic Stumpf, and Georg Sigl. Localized electromagnetic analysis of ro PUFs. In *2013 IEEE International Symposium on Hardware-Oriented Security and Trust (HOST)*, pages 19–24. IEEE, 2013.

113. Daniel Miessler. Hp study reveals 70 percent of Internet of Things devices vulnerable to attack, Mar 2015.

114. Takuji Miki, Makoto Nagata, Hiroki Sonoda, Noriyuki Miura, Takaaki Okidono, Yuuki Araga, Naoya Watanabe, Haruo Shimamoto, and Katsuya Kikuchi. A SI-backside protection circuits against physical security attacks on flip-chip devices. In *2019 IEEE Asian Solid-State Circuits Conference (A-SSCC)*, pages 25–28. IEEE, 2019.

115. Mohd Syafiq Mispan, Basel Halak, and Mark Zwolinski. NBTI aging evaluation of PUF-based differential architectures. In *2016 IEEE 22nd International Symposium on On-Line Testing and Robust System Design (IOLTS)*, pages 103–108. IEEE, 2016.

116. Aikaterini Mitrokotsa, Melanie R Rieback, and Andrew S Tanenbaum. Classification of RFID attacks. *Gen*, 15693(14443):14, 2010.

117. Maxime Montoya, Thomas Hiscock, Simone Bacles-Min, Anca Molnos, and Jacques Fournier. Adaptive masking: a dynamic trade-off between energy consumption and hardware security. In *2019 IEEE 37th International Conference on Computer Design (ICCD)*, pages 559–566. IEEE, 2019.

118. Rijoy Mukherjee and Rajat Subhra Chakraborty. Novel hardware trojan attack on activation parameters of FPGA-based DNN accelerators. *IEEE Embedded Systems Letters*, 2022.

119. Phuong Ha Nguyen, Durga Prasad Sahoo, Chenglu Jin, Kaleel Mahmood, Ulrich Rührmair, and Marten van Dijk. The interpose PUF: Secure PUF design against state-of-the-art machine learning attacks. *IACR Transactions on Cryptographic Hardware and Embedded Systems*, pages 243–290, 2019.

120. Fabian Oboril and Mehdi B Tahoori. Extratime: Modeling and analysis of wearout due to transistor aging at microarchitecture-level. In *IEEE/IFIP International Conference on Dependable Systems and Networks (DSN 2012)*, pages 1–12. IEEE, 2012.

121. Tolulope A Odetola and Syed Rafay Hasan. Sowaf: Shuffling of weights and feature maps: A novel hardware intrinsic attack (HIA) on convolutional neural network (CNN). In *2021 IEEE International Symposium on Circuits and Systems (ISCAS)*, pages 1–5. IEEE, 2021.

122. Christof Paar and Jan Pelzl. *Understanding cryptography: a textbook for students and practitioners*. Springer Science & Business Media, 2009.

123. Eric Peeters, François-Xavier Standaert, and Jean-Jacques Quisquater. Power and electromagnetic analysis: Improved model, consequences and comparisons. *Integration*, 40(1):52–60, 2007.

124. Stevan Preradovic, Nemai C. Karmakar, and Isaac Balbin. RFID transponders. *IEEE Microwave Magazine*, 9(5):90–103, 2008.

125. Lennart M Reimann, Luca Hanel, Dominik Sisejkovic, Farhad Merchant, and Rainer Leupers. Qflow: Quantitative information flow for security-aware hardware design in verilog. In *2021 IEEE 39th International Conference on Computer Design (ICCD)*, pages 603–607. IEEE, 2021.

126. Martin Riedmiller and Heinrich Braun. A direct adaptive method for faster backpropagation learning: The RPROP algorithm. In *IEEE international conference on neural networks*, pages 586–591. IEEE, 1993.

127. Ronald L Rivest, Adi Shamir, and Leonard Adleman. A method for obtaining digital signatures and public-key cryptosystems. *Communications of the ACM*, 21(2):120–126, 1978.

128. Pankaj Rohatgi. Electromagnetic attacks and countermeasures. *Cryptographic Engineering*, pages 407–430, 2009.

129. Masoud Rostami, Mehrdad Majzoobi, Farinaz Koushanfar, Daniel S Wallach, and Srinivas Devadas. PUF authentication and key-exchange by substring matching, April 18 2017. US Patent 9,628,272.

130. Jarrod A Roy, Farinaz Koushanfar, and Igor L Markov. Epic: Ending piracy of integrated circuits. In *Proceedings of the conference on Design, automation and test in Europe*, pages 1069–1074, 2008.

131. Ulrich Rührmair, Jan Sölter, Frank Sehnke, Xiaolin Xu, Ahmed Mahmoud, Vera Stoyanova, Gideon Dror, Jürgen Schmidhuber, Wayne Burleson, and Srinivas Devadas. PUF modeling attacks on simulated and silicon data. *IEEE transactions on information forensics and security*, 8(11):1876–1891, 2013.

132. Ulrich Rührmair, Jan Sölter, Frank Sehnke, Xiaolin Xu, Ahmed Mahmoud, Vera Stoyanova, Gideon Dror, Jürgen Schmidhuber, Wayne Burleson, and Srinivas Devadas. PUF modeling attacks on simulated and silicon data. *IEEE Transactions on Information Forensics and Security*, 8(11):1876–1891, 2013.

133. Ahmad-Reza Sadeghi, Ivan Visconti, and Christian Wachsmann. PUF-enhanced RFID security and privacy. In *Workshop on secure component and system identification (SECSI)*, volume 110, 2010.

134. Ahmad-Reza Sadeghi, Ivan Visconti, and Christian Wachsmann. PUF-enhanced RFID security and privacy. In *Workshop on secure component and system identification (SECSI)*, volume 110, 2010.

135. Durga Prasad Sahoo, Debdeep Mukhopadhyay, Rajat Subhra Chakraborty, and Phuong Ha Nguyen. A multiplexer-based arbiter PUF composition with enhanced reliability and security. *IEEE Transactions on Computers*, 67(3):403–417, 2017.

136. Pranesh Santikellur, Aritra Bhattacharyay, and Rajat Subhra Chakraborty. Deep learning based model building attacks on arbiter PUF compositions. *IACR Cryptol. ePrint Arch.*, 2019:566, 2019.

137. Sanjay E Sarma. Towards the five-cent tag (technical report mit-AUTOID-WH-006), 2001.

138. Alexander Schlösser, Dmitry Nedospasov, Juliane Krämer, Susanna Orlic, and Jean-Pierre Seifert. Simple photonic emission analysis of AES. In *International Workshop on Cryptographic Hardware and Embedded Systems*, pages 41–57. Springer, 2012.

139. Pete Sedcole and Peter YK Cheung. Within-die delay variability in 90 nm FPGAS and beyond. In *2006 IEEE International Conference on Field Programmable Technology*, pages 97–104. IEEE, 2006.

140. Pete Sedcole and Peter YK Cheung. Within-die delay variability in 90 nm FPGAS and beyond. In *2006 IEEE International Conference on Field Programmable Technology*, pages 97–104. IEEE, 2006.

141. Mohamed Seliem, Khalid Elgazzar, and Kasem Khalil. Towards privacy preserving IoT environments: a survey. *Wireless Communications and Mobile Computing*, 2018:1–15, 2018.

142. Kaveh Shamsi, Meng Li, Kenneth Plaks, Saverio Fazzari, David Z Pan, and Yier Jin. Ip protection and supply chain security through logic obfuscation: A systematic overview. *ACM Transactions on Design Automation of Electronic Systems (TODAES)*, 24(6):1–36, 2019.

143. James E Stine, Ivan Castellanos, Michael Wood, Jeff Henson, Fred Love, W Rhett Davis, Paul D Franzon, Michael Bucher, Sunil Basavarajaiah, Julie Oh, et al. Freepdk: An open-source variation-aware design kit. In *2007 IEEE international conference on Microelectronic Systems Education (MSE'07)*, pages 173–174. IEEE, 2007.

144. James E Stine, Ivan Castellanos, Michael Wood, Jeff Henson, Fred Love, W Rhett Davis, Paul D Franzon, Michael Bucher, Sunil Basavarajaiah, Julie Oh, et al. Freepdk: An open-source variation-aware design kit. In *2007 IEEE international conference on Microelectronic Systems Education (MSE'07)*, pages 173–174. IEEE, 2007.

145. Cynthia Sturton, Matthew Hicks, David Wagner, and Samuel T King. Defeating UCI: Building stealthy and malicious hardware. In *2011 IEEE symposium on security and privacy*, pages 64–77. IEEE, 2011.

146. G Edward Suh and Srinivas Devadas. Physical unclonable functions for device authentication and secret key generation. In *2007 44th ACM/IEEE Design Automation Conference*, pages 9–14. IEEE, 2007.

147. G Edward Suh and Srinivas Devadas. Physical unclonable functions for device authentication and secret key generation. In *2007 44th ACM/IEEE Design Automation Conference*, pages 9–14. IEEE, 2007.

148. Berk Sunar, William J Martin, and Douglas R Stinson. A provably secure true random number generator with built-in tolerance to active attacks. *IEEE Transactions on computers*, 56(1):109–119, 2007.

149. Koyo Suzuki, Katsuyoshi Miura, and Koji Nakamae. NBTI/PBTI tolerant arbiter PUF circuits. In *2017 IEEE 23rd International Symposium on On-Line Testing and Robust System Design (IOLTS)*, pages 80–84. IEEE, 2017.

150. Mohammad Tehranipoor and Farinaz Koushanfar. A survey of hardware trojan taxonomy and detection. *IEEE design & test of computers*, 27(1):10–25, 2010.

151. Ferucio Laurenţiu Ţiplea, Cristian Andriesei, and Cristian Hristea. Security and privacy of PUF-based RFID systems. In *Cryptography-Recent Advances and Future Developments*. IntechOpen, 2020.

152. Johannes Tobisch and Georg T Becker. On the scaling of machine learning attacks on PUFs with application to noise bifurcation. In *International Workshop on Radio Frequency Identification: Security and Privacy Issues*, pages 17–31. Springer, 2015.

153. Johannes Tobisch and Georg T Becker. On the scaling of machine learning attacks on PUFs with application to noise bifurcation. In *International Workshop on Radio Frequency Identification: Security and Privacy Issues*, pages 17–31. Springer, 2015.

154. Anthony Van Herrewege, Stefan Katzenbeisser, Roel Maes, Roel Peeters, Ahmad-Reza Sadeghi, Ingrid Verbauwhede, and Christian Wachsmann. Reverse fuzzy extractors: Enabling lightweight mutual authentication for PUF-enabled RFIDs. In *International Conference on Financial Cryptography and Data Security*, pages 374–389. Springer, 2012.

155. Nidish Vashistha, M Tanjidur Rahman, Olivia P Dizon-Paradis, and Navid Asadizanjani. Is backside the new backdoor in modern socs? In *2019 IEEE International Test Conference (ITC)*, pages 1–10. IEEE, 2019.

156. Wei-Che Wang, Yair Yona, Suhas Diggavi, and Puneet Gupta. Ledpuf: Stability-guaranteed physical unclonable functions through locally enhanced defectivity. In *2016 IEEE International Symposium on Hardware Oriented Security and Trust (HOST)*, pages 25–30. IEEE, 2016.

157. Zheng Wang, Yi Chen, Aakash Patil, Jayasanker Jayabalan, Xueyong Zhang, Chip-Hong Chang, and Arindam Basu. Current mirror array: A novel circuit topology for combining physical unclonable function and machine learning. *IEEE Transactions on Circuits and Systems I: Regular Papers*, 65(4):1314–1326, 2017.

158. Stephen A Weis. RFID (radio frequency identification): Principles and applications. *System*, 2(3):1–23, 2007.

159. Yuejiang Wen and Yingjie Lao. Enhancing PUF reliability by machine learning. In *2017 IEEE International Symposium on Circuits and Systems (ISCAS)*, pages 1–4. IEEE, 2017.

160. Meng-Yi Wu, Tsao-Hsin Yang, Lun-Chun Chen, Chi-Chang Lin, Hao-Chun Hu, Fang-Ying Su, Chih-Min Wang, James Po-Hao Huang, Hsin-Ming Chen, Chris Chun-Hung Lu, et al. A PUF scheme using competing oxide rupture with bit error rate approaching zero. In *2018 IEEE International Solid-State Circuits Conference-(ISSCC)*, pages 130–132. IEEE, 2018.

161. Chengjie Xi, Nathan Jessurun, and Navid Asadizanjani. A framework to assess the security of advanced integrated circuit (IC) packaging. In *2020 IEEE 8th Electronics System-Integration Technology Conference (ESTC)*, pages 1–7. IEEE, 2020.

162. Xiaolin Xu, Wayne Burleson, and Daniel E Holcomb. Using statistical models to improve the reliability of delay-based PUFs. In *2016 IEEE Computer Society Annual Symposium on VLSI (ISVLSI)*, pages 547–552. IEEE, 2016.

163. Xiaolin Xu and Daniel Holcomb. A clockless sequential PUF with autonomous majority voting. In *Proceedings of the 26th edition on Great Lakes Symposium on VLSI*, pages 27–32, 2016.

164. Jeongkyu Yang. Security and privacy on authentication protocol for low-cost radio frequency identification. *A Thesis for the Degree of Master of Science*, 2005.

165. Meng-Day Yu and Srinivas Devadas. Secure and robust error correction for physical unclonable functions. *IEEE Design & Test of Computers*, 27(1):48–65, 2010.

166. Meng-Day Yu, Matthias Hiller, Jeroen Delvaux, Richard Sowell, Srinivas Devadas, and Ingrid Verbauwhede. A lockdown technique to prevent machine learning on PUFs for lightweight authentication. *IEEE Transactions on Multi-Scale Computing Systems*, 2(3):146–159, 2016.

167. Meng-Day Yu, Matthias Hiller, Jeroen Delvaux, Richard Sowell, Srinivas Devadas, and Ingrid Verbauwhede. A lockdown technique to prevent machine learning on PUFs for lightweight authentication. *IEEE Transactions on Multi-Scale Computing Systems*, 2(3):146–159, 2016.

168. Meng-Day Yu, David M'Raïhi, Ingrid Verbauwhede, and Srinivas Devadas. A noise bifurcation architecture for linear additive physical functions. In *2014 IEEE International Symposium on Hardware-Oriented Security and Trust (HOST)*, pages 124–129. IEEE, 2014.

169. Meng-Day Yu, David M'Raïhi, Ingrid Verbauwhede, and Srinivas Devadas. A noise bifurcation architecture for linear additive physical functions. In *2014 IEEE International Symposium on Hardware-Oriented Security and Trust (HOST)*, pages 124–129. IEEE, 2014.

170. Siarhei S Zalivaka, Alexander A Ivaniuk, and Chip-Hong Chang. Reliable and modeling attack resistant authentication of arbiter PUF in FPGA implementation with trinary quadruple response. *IEEE Transactions on Information Forensics and Security*, 14(4):1109–1123, 2018.

171. Chen Zhou, Saroj Satapathy, Yingjie Lao, Keshab K Parhi, and Chris H Kim. Soft response generation and thresholding strategies for linear and feed-forward MUX PUFs. In *Proceedings of the 2016 International Symposium on Low Power Electronics and Design*, pages 124–129, 2016.